PLANT SYSTEMATICS

PLANT SYSTEMATICS

GURCHARAN SINGH

Science Publishers, Inc.
U.S.A.

SCIENCE PUBLISHERS, Inc.
Post Office Box 699
Enfield, New Hampshire 03748
United States of America

Internet site: *http://www.scipub.net*

sales@scipub.net (marketing department)
editor@scipub.net (editorial department)
info@scipub.net (for all other enquiries)

Library of Congress Cataloging-in-Publication Data

Singh, Gurcharan.
   Plant systematics/Gurcharan Singh.
     p.cm.
   Includes bibliographical references (p.).
   ISBN 1-57808-081-9 (hc.) - ISBN 1-57808-077-0 (pbk.)
   1. Botany-Classification, I. Title.

QK95 .S57 1999
580'.1'2-dc21
                                                    99-048015

ISBN 1-57808-077-0 (Paperback)
ISBN 1-57808-081-9 (Hardcover)

Published by Science Publishers Inc., Enfield, NH, USA
Printed in India.

Preface

Plant systematics is a fast developing field but unfortunately with very few and occasional titles. Frequent revisions in major classification systems, development of new tools, methodology and concepts send the older books into oblivion far too soon. *Taxonomy of Vascular Plants* by Lawrence (1951), which was a rage in the fifties and early sixties was largely replaced by the book *Principles of Angiosperm Taxonomy* by Davis and Heywood (1963). The latter continued its domination in the sixties and seventies until two important publications were simultaneously presented (luckily with different emphasis on selective topics) by Jones and Luchsinger (1979, *Plant Systematics*) and Stace (1980, *Plant Taxonomy and Biosystematics*). They have published revisions in 1986 and 1989 respectively with some improvements. More recently, Woodland (1991, *Contemporary Plant Systematics*) produced updated information in the field. The frequent revisions of contemporary systems of classification by Takhtajan (1983, 1987, 1997), Dahlgren (1989) and Thorne (1992, 1997 electronic version) have outpaced these textbooks.

While writing a book on plant systematics, one has to grapple with the temptation to include descriptions of various families (which often take up more than 50% of the total content, but are so easily available in so many books, with the information scarcely changing). Those who have managed to suppress this temptation have been able to present valuable books on systematics, as for example Davis and Heywood. Stace maintained the same format but included limited information on fundamental aspects of systematics (identification, nomenclature, phylogeny) with an overemphasis on experimental studies. There is also general brevity of information to maintain a smaller size of the book. Authors such as Porter (1959, *Taxonomy of Flowering Plants*), Jones and Luchsinger, Radford (1986, *Fundamentals of Plant Systematics*), and Woodland, who preferred to include description of families (though selected ones) have done so at the expense of general taxonomic information and with increased book volume. This consideration has been instrumental in the decision to keep out the description of families in the present volume, and to devote a balanced attention to various aspects of plant systematics.

The author has attempted to strike a balance between classical fundamental information and recent developments in plant systematics. Special attention has been devoted to information on botanical nomenclature, identification and phylogeny of angiosperms with numerous relevant examples and detailed explanation of important nomenclatural problems. An attempt has been made to present a continuity between orthodox and contemporary identification methods by working on a common example.

Internet highways are revolutionising the exchange of scientific information. Botanical organisations have plunged into this revolution in a big way. Instant information on major classification systems, databases, herbaria, gardens, indices and thousands of illustrations are available to users worldwide at the touch of a button. Discussion on important aspects of this information highway and useful links to this information are provided at various places in this book, with a special chapter on the Web information. My interaction with colleagues, especially the members of the list 'Taxacom', has been greatly helpful in resolving some doubts. I am thankful to all respondents whose information has been valuable on several occasions.

My experience of nearly three decades of teaching has prompted me to simplify the explanation of the concepts regarding phylogeny without compromising the information content. A similar approach has been followed while providing information on rather complicated methods of numerical taxonomy such as cluster analysis and ordination as also the newer and rather conflicting concepts about cladistics.

By not including information on representative families, I may have disappointed some readers, but this decision to exclude family descriptions has enabled me to present information on various other aspects of systematics.

For providing me inspiration for this book, I am indebted to my undergraduate students, who helped me to improve the material through frequent interactions. I am also indebted to my wife Mrs. K.G. Singh for constantly bearing with my overindulgence with this book and also providing ready help in computer-related activities during the processing of this work.

I wish to record thanks to all the colleagues whose inputs have helped me to improve the information presented here. I also place on record sincere thanks to Dr. Jef Veldkamp for valuable information on nomenclature, Dr. Gertrud Dahlgren for photographs and literature concerning her and her husbands' work. I am also thankful to the authorities of the New York Botanical Garden for photographs used in the book, Royal Botanical Garden, Kew and Cambridge University Press for permission to reproduce figures.

New Delhi   GURCHARAN SINGH
September, 1999

Contents

Taxonomy and Systematics

Taxonomy is basically concerned with the classification of organisms. Plants are man's prime companions in this universe, being the source of food and energy, shelter and clothing, drugs and beverages, oxygen and aesthetic environment. Humans learn to identify, describe, name and classify foods, clothes, books, games, vehicles, religions, professions and any other objects that they come across or that influence their life. The process begins and ends with life. Each member of *Homo sapiens* is as such a taxonomist from cradle to grave. Classification is thus a natural occupation of man, a recognition of one of his most necessary pastimes (Davis and Heywood, 1963). Although man has been classifying plants since the advent of civilisation, **taxonomy** was recognised as a formal subject only in 1813 by A.P. de Candolle as a combination of **taxis** (arrangement) and **nomos** (rules or laws).

For a long time plant taxonomy was considered as 'the science of identifying, naming and classifying plants' (Lawrence, 1951). Since identification and nomenclature are important prerequisites for any classification, taxonomy is often defined as the 'science dealing with the study of classification, including its bases, principles, rules and procedures' (Davis and Heywood, 1963).

Systematics was initially recognised as a more inclusive field of study concerned with the diversity of plants and their naming, classification and evolution. Simpson (1961) defined systematics as a 'scientific study of kinds and diversity of organisms, and of any and all relationships between them'. The scope of taxonomy has, however, been enlarged in recent years to make taxonomy and systematics synonymous. Some authors prefer to differentiate between them, giving systematics a broader definition and restricting taxonomy to the study of classification. A broader definition (Stace, 1980) of **taxonomy** to coincide with **systematics** recognises it as 'the study and description of variation in organisms, the investigation of causes and consequences of this variation, and the manipulation of the data obtained to produce a system of classification'. Taxonomy now is thus a broadened field meant for the study of nomenclature, description, classification, identification and phylogeny.

Prior to the evolutionary theory of Darwin, relationships were expressed as **natural affinities** on the basis of overall similarity in morphological features. Darwin ushered in an era of assessing **phylogenetic relationships** based on the course of evolutionary descent. With the introduction of computers and refined statistical procedures overall similarity is represented as **phenetic relationship**, which takes into account every available feature, derived from such diverse fields as anatomy, embryology, morphology, palynology, cytology, phytochemistry, physiology, ecology, phytogeography and ultrastructure. With the advancement of biological fields, new information flows continuously and the taxonomists are faced with the challenge of integrating and providing a synthesis of all available information. Taxonomy now is thus an **unending synthesis**, a dynamic science with never-ending duties. The continuous flow of data necessitates rendering descriptive information, revising schemes of identification, re-evaluating, and improving systems of classification and perceiving new relationships for understanding plants. The discipline as such includes all activities that are part of the effort to organise and record diversity of plants and appreciate the fascinating differences among the species of plants. Taxonomic activities are basic to all other biological sciences, but also depend, in turn, on other disciplines for data and information useful in constructing classification.

I. BASIC COMPONENTS OF SYSTEMATICS

Various systematic activities are directed towards the singular goal of constructing an ideal system of classification that necessitates the procedures of identification, description, nomenclature and constructing affinities.

A. Classification

Classification is an arrangement of plants into groups on the basis of similarities. The groups are in turn assembled into more inclusive groups, until all have been assembled into a single most inclusive group. In sequence of increasing inclusiveness groups are assigned to a fixed hierarchy of categories such as species, genus, family, order, class and division, the final arrangement constituting a system of classification. The process of classification includes assigning **appropriate position and rank** to a new **taxon** (a taxonomic group assigned to any rank; pl. **taxa**), **dividing** a taxon into smaller units, **uniting** two or more taxa into one, **transferring** its position from one group to another and **altering** its rank. Once established, a classification provides an important mechanism of information storage, retrieval and usage. Taxonomic entities are classified in different fashions:

1. **Artificial classification** is utilitarian, based on arbitrary, easily observable characters such as habit, colour, number, form or similar features. The **sexual system** of Linnaeus, which fits in this category, utilised the number of stamens for primary classification of flowering plants.

2. **Natural classification** uses overall similarity in grouping taxa, a concept initiated by M. Adanson and culminating in the extensively used classification of Bentham and Hooker. Natural systems of 18th and 19th centuries used morphology in delimiting overall similarity. The concept of overall similarity has undergone considerable refinement in recent years. As against the sole morphological features as indicators of similarity in natural systems, overall similarity is now judged based on features derived from morphology, anatomy, embryology, phytochemistry, ultrastructure and in fact all other fields of study, an approach which has substituted **phenetic relationship** as a more appropriate term to define overall similarity. Phenetic relationship is very prominently used in modern phylogenetic systems to decide realignments within the system of classification.

3. **Phylogenetic classification** is based on the evolutionary descent of a group of organisms, the relationship depicted through a **phylogram**, **phylogenetic tree** or a **cladogram**. Classification is constructed with this premise in mind, that all descendants of a common ancestor should be placed in the same group (i.e., group should be **monophyletic**). An **evolutionary taxonomic classification**, a disparate philosophy of classification, accepts leaving out certain descendants of a common ancestor (i.e., recognising **paraphyletic** groups), thus failing to provide a true picture of geological history.

The contemporary phylogenetic systems of classification, including those of Takhtajan, Cronquist, Thorne and Dahlgren, are largely based on decisions in which **phenetic information** is liberally used in deciding the phylogenetic relationship between groups, differing largely on the weightage given to the cladistic or phenetic relationship.

B. Identification

Identification or determination is recognising an unknown specimen with an already known taxon, and assigning a correct rank and position in an extant classification. In practice, it involves finding a name for an unknown specimen using identified herbarium specimens, dichotomous keys, polyclaves or computer-aided devices.

C. Description

The description of a taxon involves listing its features by recording appropriate character states. A shortened description consisting of only those taxonomic characters which help separate a taxon from closely related taxa, forms **diagnosis**, and the characters are 'ermed **diagnostic characters**. The diagnostic characters for a taxon determine its **circumscription**. Description is recorded in a set pattern (habit, stem, leaves, flower, sepals, petals, stamens, carpels, fruit etc.). For each character an appropriate character state is listed. Flower colour (character) may thus be red, yellow, white etc. (states).

D. Nomenclature

Nomenclature deals with determination of a **correct name** for a taxon using rules and recommendations of the **International Code of Botanical Nomenclature (ICBN)**. Updated every six years or so, the **Code** helps in picking up a single correct name out of numerous scientific names available for a taxon, with a particular circumscription, position and rank. To avoid inconvenient name changes for certain taxa, a list of conserved names is provided in the Code.

E. Phylogeny

Phylogeny is the study of the genealogy and evolutionary history of a taxonomic group. Genealogy is the study of ancestral relationships and lineages. Relationships are depicted through a diagram better known as a **phylogram** (Stace, 1989) since the commonly used term **cladogram** is more appropriately used for a diagram constructed through cladistic methodology. A phylogram is a branching diagram based on the degree of advancement (**apomorphy**) in the descendants, the longest branch representing the most advanced group. It is distinct from **a phylogenetic tree** in which the vertical scale represents a geological time-scale and all living groups reach the top, with primitive ones near the centre and advanced ones near the periphery. Monophyletic groups, including all descendants of a common ancestor, are recognised and form entities in a classification system. Paraphyletic groups in which some descendants of a common ancestor are left out are reunited. Polyphyletic groups, with more than one common ancestor, are split to form monophyletic groups. Phenetic information may often help in determining a phylogenetic relationship.

II. AIMS OF SYSTEMATICS

The activities of plant systematics are basic to all other biological sciences and, in turn, depend on these for additional information useful in constructing a classification. These activities are directed towards achieving the undermentioned aims:

1. To provide a convenient method of identification and communication. A workable classification having taxa arranged in hierarchy, detailed and diagnostic descriptions are essential for identification. Properly identified and arranged herbarium specimens, dichotomous keys,

polyclaves and computer-aided identification are important aids for identification. The Code (ICBN), written and documented through the efforts of IAPT (International Association of Plant Taxonomy), helps in deciding the single correct name acceptable to the whole botanical community.

2. To provide an inventory of the World's flora. Though a single World flora is difficult to come by, floristic records of continents (**Continental Floras;** *cf. Flora Europaea* by Tutin *et al.*), regions or countries (**Regional Floras;** *cf. Flora of British India* by J. D. Hooker) and states or counties (**Local Floras;** *cf. Flora of Delhi* by J. K. Maheshwari) are well documented. In addition, **World Monographs** for selected genera (e.g., *The genus Crepis* by Babcock) and families (e.g., *Das pflanzenreich* ed. by A. Engler) are also available.

3. To detect evolution at work. To reconstruct the evolutionary history of the plant kingdom, determining the sequence of evolutionary change and character modification.

4. To provide a system of classification which depicts evolution within the group. The evolutionary relationship between the groups is commonly depicted with the help of a phylogram wherein the longest branches represent more advanced groups and the shorter, nearer the base, primitive ones. In addition, the groups are represented by balloons of different sizes proportional to the number of species in the respective groups. Such a phylogram is popularly known as a **bubble diagram**.

5. To provide an integration of all available information. To gather information from all fields of study, analyse this information using statistical procedures with the help of computers, provide a synthesis of this information and develop a classification based on overall similarity. This synthesis is unending, however, since scientific progress will continue and new information will continue to develop and pose new challenges for taxonomists.

6. To provide an information reference, supplying methodology for information storage, retrieval, exchange and utilisation. To provide significantly valuable information concerning endangered species, unique elements, genetic and ecological diversity.

7. To provide new concepts, reinterpret the old, and develop new procedures for correct determination of taxonomic affinities, in terms of phylogeny and phenetics.

III. ADVANCEMENT LEVELS IN SYSTEMATICS

Plant systematics has made considerable strides from herbarium records to databanks recording information on every possible attribute of a plant. Because of extreme climatic diversity, floristic variability, inaccessibility of certain regions and economic disparity of different regions, the present-day taxonomy finds itself in different stages of advancement in different parts of the world. This can conveniently be divided into four distinct **phases** encountered in different parts of the world today:

1. **Exploratory or Pioneer phase:** This is the beginning of plant taxonomy, collecting specimens and building herbarium records. The few specimens of a species in the herbarium are the only record of its variation. These specimens are, however, useful in preliminary inventory of flora through discovery, description, naming and identification of plants. Most areas of tropical Africa and tropical Asia experience this phase.

2. **Consolidation or Systematic phase:** During this phase herbarium records are ample and enough information is available concerning variation from field studies. This development is helpful in the preparation of Floras and Monographs. Most parts of southern Europe experience this phase.

3. **Experimental or Biosystematic phase:** During this phase the herbarium records and variation studies are complete. In addition information on **biosystematics** (studies on transplant experiments, breeding behaviour and chromosomes) is also available. Central Europe has reached this phase of plant taxonomy.

4. **Encyclopaedic or Holotaxonomic phase:** Here not only the previous three phases are attained, but information on all botanical fields is also available. This information is assembled, analysed, and a meaningful synthesis of analysis is provided for understanding phylogeny. Collection of data, analysis and synthesis are the jobs of an independent discipline of taxonomy, **numerical taxonomy**.

The first two phases of taxonomy are often considered under **alpha-taxonomy** and the last phase under **omega-taxonomy**. At present only a few persons are involved in encyclopaedic work and that, too, in a few isolated taxa. It may thus be safe to conclude that though in a few groups omega-taxonomy is within reach, for the great majority of plants, mainly in the tropics, the 'alpha' stage has not been crossed. The total integration of available information for the plant kingdom is thus only a distant dream at present.

The diversification of plant taxonomy or systematics in recent years has resulted in the identification of distinct disciplines. **Biosystematics** is the field of study dealing with transplant experiments, breeding behaviour, chromosomes and variation in plants. **Chemotaxonomy** utilises chemical evidence in solving taxonomic problems. **Phylosystematics (phylogenetics)** deals with phylogeny in classification. **Numerical taxonomy** or **taximetrics** uses numerical methods for analysis and synthesis of information derived from various fields, attaining grouping based on a **phenetic relationship**. More recently cladistics is developing as a prominent field of study based on phylogenetic systematics; it has developed a distinct methodology to analyse phylogenetic data on lines parallel to taximetrics, but whereas taximetric studies finally result in the construction of a **phenogram**, the cladistic methodology constructs a **cladogram**.

IV. SYSTEMATICS IN INTERNET REVOLUTION

Electronic revolution in the last few years and the development of Internet sites has seen taxonomy becoming a dominant component of Botany on the web. The three major contemporary systems of angiosperm classification, with their latest versions—**Cronquist** (1988), **Takhtajan** (1997) and **Thorne** (1992)—are already on the web with links to various groups recognised by these authors. Thorne's electronic version (1997) with promises of periodic upgradation contains considerable changes from his printed version (1992). Such an approach will go a long way to maintaining communication of new ideas with minimised time gap. The latest version of the **Botanical Code** (Tokyo Code, 1994) with appropriate links to various articles is available for consultation. A **Draft BioCode,** an attempt at a unified Code for all living organisms, is already on the web for discussion among biologists and for possible adoption in the first decade of the next century. Internet is also helping in a big way to maintain servers with searchable software for locating taxonomists, herbaria, books, journals, gardens, families, genera, plant names and a host of other taxonomic components. Besides **Lists** and **News Groups** maintain a platform of active interaction among taxonomists. Perhaps no branch of botany today is so prominent on the web as taxonomy. Most of the prominent Institutions, Botanical Gardens, Publishers and Associations are already on the Internet with servers, which afford ready access to their literature, staff and research activities.

2

Historical Background of Plant Classification

The urge to classify plants has been with man since he first set his foot on this planet, borne of a need to know what he should eat, avoid, use as cures for ailments and utilise for his shelter. Initially this information was accumulated and stored in the human brain and passed on to generations through word of mouth in dialects restricted to small communities. Slowly man learnt to put his knowledge in black and white for others to share and improve. We have now reached a stage whereby a vast amount of information can be conveniently stored and utilised for far-reaching conclusions aimed at developing ideal systems of classification, which depict putative relationships between organisms. Historical development of classification has passed through four distinct approaches, beginning with simple classifications based on gross morphology to the latest phylogenetic systems incorporating all phenetic information.

1. CLASSIFICATIONS BASED ON GROSS MORPHOLOGY

Classifications based on features studied without microscopic aids continued until the 17th century when the naked eye was the sole tool of observation. The trail backwards leads us to preliterate man.

A. Preliterate Mankind

Although no written records of the activities of our preliterate ancestors are available, it is safe to assume that they were practical taxonomists knowing which plants were edible and which cured their ailments. Primitive tribes in remote areas of the world still carry the tradition of preserving knowledge of the names and uses of plants by word of mouth from one generation to another. Such classifications of plants developed by isolated communities through societal need and without the influence of science are termed **Folk taxonomies;** they often parallel modern taxonomy. The common English names grass and sedge are equivalent to the modern families Poaceae and Cyperaceae and illustrate this parallel development between Folk taxonomy and Modern taxonomy.

B. Early Literate Civilisations

Early civilisations flourished in Babylonia, Egypt, China and India. Though the written records of Indian botany appeared several centuries before those of the Greeks, they remained in obscurity, not reaching the outside world. Moreover they were written in Sanskrit, a language not easily understood in the West. Crops such as wheat, barley, dates, melons and cotton were grown during the Vedic Period (2000 B.C. to 800 B.C.). Indians obviously knew about descriptive botany and cultural practices. The first world symposium on medicinal plants was held in the Himalayan region in the 7th century B.C.

Theophrastus (370 B.C. to 285 B.C.)—**Father of botany**. Theophrastus was a disciple of Plato and later Aristotle, and after the latter's death he inherited his library and his garden. He rose to become the head of the Lyceum at Athens. Theophrastus is credited with more than 200 works but most survive as fragments or as quotations in the works of other authors. Two of his botanical works have survived intact, however, and are available in English translations: *Enquiry into Plants* (1916) and *The Causes of Plants* (1927). He described about 500 kinds of plants, classified into four major groups: trees, shrubs, subshrubs and herbs. He recognised differences between flowering plants and non-flowering plants, superior ovary and inferior ovary, free and fused petals and fruit types. He was aware of the fact that many cultivated plants do not breed true. Several names used by Theophrastus in his *De Historia plantarum*, e.g. *Daucus, Crataegus* and *Narcissus*, to name a few, are in use even today.

Theophrastus was fortunate to have Alexander the Great as his patron. During his conquests Alexander made arrangements to send back materials to Athens enabling Theophrastus to write about exotic plants such as cotton, cinnamon and bananas. Botanical knowledge at the Lyceum in Athens thus flourished during this truly golden age of learning whose botanical advance Theophrastus was privileged to steer.

Parasara (250 B.C.to 120 B.C.)—**Indian scholar.** Parasara was an Indian scholar who compiled *Vrikshayurveda* (Science of Plant Life), one of the earliest works dealing with plant life from a scientific standpoint, a manuscript discovered a few decades ago. The book has separate chapters on morphology, properties of soil, forest types of India and details of internal structure, which suggest that he had a magnifying apparatus of some kind.

He also described the existence of cells (rasakosa) in the leaf, transportation of soil solution from the root to the leaves where it is digested by means of chlorophyll [ranjakena pacyamanat] becoming a nutritive substance and yielding by-products. Plants were classified into numerous families [ganas] on the basis of morphological features not known to the European classification until the 18th century. Samiganyan [Leguminosae] were distinguished by hypogynous flowers, five petals of different sizes, gamosepalous calyx and a fruit, actually a legume. Svastikaganyan [Cruciferae] similarly has a calyx resembling a swastika, ovary superior, 4 free sepals and petals each, six stamens, of which 2 are shorter, and 2 carpels forming a bilocular fruit. Unfortunately, this great scientific advance did not reach Europe at that time where scientific knowledge was just making its debut.

Among the other Indian scholars, **Caraka (Charaka**—Ist century A.D.) wrote *Caraka Samhita (Charaka Samhita)*, in which he recognised trees without flowers, trees with flowers, herbs which wither after fructification and other herbs with spreading stems as separate groups. This huge treatise on Indian medicine, containing eight divisions, is largely based on a much earlier treatise published by Agnivesh. A.C. Kaviratna translated it into English in 1897.

Caius Plinius Secundus (23 A.D. to 79 A.D.)—Pliny the Elder.

The decline of the Greek Empire saw the emergence of the Romans. Pliny, a naturalist who served under the Roman army, attempted a compilation of everything known about the world in an extensive 37-volume work *Historia naturalis*, 9 volumes of which were devoted to medicinal plants. In spite of a few errors and fanciful tales from travellers, the Europeans held this work in reverential awe for many centuries. He died during the eruption of Vesuvius.

Pedanios Dioscorides (1st century A.D.).

Dioscorides, of Greek parentage, was a native of the Roman province Cicilia. Being a physician in the Roman army, he travelled extensively and gained first-hand knowledge about plants used in treating various ailments. He wrote a truly outstanding work, *Materia medica*, presenting an account of nearly 600 medicinal plants, nearly 100 more than Theophrastus. Excellent illustrations were added later. Written in a straightforward style, the book was an asset for any literate man for the next 15 centuries. No drug was recognised as genuine unless mentioned in *Materia medica*. It is no less a tribute to Dioscorides that a beautiful illustrated copy of the book was prepared for Emperor Flavius Olybrius Anicius around 500 A.D., who presented it as a gift to his beautiful daughter, Princess Juliana Anicia. The manuscript, better known as *Codex Juliana*, is a prize preserved in Vienna. *Materia medica* was not a deliberate attempt at classification but legumes, mints and umbels were described as separate groups.

C. Medieval Botany

During the Middle Ages (5th to 15th century A.D.) little or no progress was made in botanical investigation. During this dark period in history Europe and Asia witnessed wars, famine and epidemics, and the only worthwhile contribution was copying and recopying of earlier manuscripts, unfortunately often with errors added. The strawberry plant was thus shown to have five leaflets instead of three in several manuscripts. The manuscripts were lost at a faster rate than they could be copied.

Islamic Botany. The ascent of the Muslim empire between 610-1100 A.D. saw the revival of literacy. Greek manuscripts were translated and preserved. Being practical people they concentrated on agriculture and medicine and produced lists of drug plants. Ibn-Sina, better known as Avicenna, authored *Canon of Medicine*, a scientific classic like *Materia medica*. Another muslim scholar, Ibu-al-Awwan, 12th century, described about 600 plants, interpreted their sexuality as well as the role of insects in fig pollination.

Albertus Magnus (ca 1193-1280 A.D.)—**Doctor Universalis**. Albertus Magnus, called **Doctor Universalis** by his contemporaries and **Aristotle of the Middle Ages** by historians, is the best-remembered naturalist of that period. He wrote on many subjects. The botanical work *De vegetabilis* dealt with medicinal plants and provided descriptions of plants based on first-hand information. He is believed to be the first to recognise monocots and dicots based on stem structure. He also separated vascular and non-vascular plants.

D. Renaissance

The 15th century saw the onset of the Renaissance in Europe, with technical innovations, mainly the printing machine and the science of navigation. Invention of the printing machine with movable type around 1440 ensured wide circulation of manuscripts. Navigation led to successful exploitation of botanical wealth in distant places.

Herbalists

Printing made books cheap. The first to become popular were medically oriented books on plants. Specialists started producing their own botanical-medical books, which were easily understood compared to ancient manuscripts. These came to be known as **herbals** and the authors who wrote these were known as **herbalists**. The first herbals were published under the name *Gart der Gesundheit* or *Hortus sanitatus*. These were cheaply done and of poor quality. The outstanding herbals came from German herbalists Otto Brunfels, Jerome Bock, Valerius Cordus and Leonard Fuchs, constituting the **German Fathers of Botany**. Otto Brunfels (1464-1534) wrote *Herbarium vivae* *eicones* in three volumes (1530-1536), a herbal that marked the beginning of modern taxonomy, and contained excellent illustrations prepared from living plants. The text, however, was of little value, being extracts from earlier writers. Jerome Bock (Hieronymus Tragus), who lived between 1498 and 1554, wrote *New kreuterbuch* in 1539, which contained no illustrations but included accurate descriptions based on first-hand knowledge, also mentioning localities and habitats. He described 567 species classified as herbs, shrubs and trees. The herbal, written in German, was widely understood compared to the manuscripts of earlier scholars, which were in Greek and Latin, languages which had by then become obsolete. Leonard Fuchs (1501-1566), regarded as more meritorious than his contemporaries, wrote *De Historia stirpium* in 1542, containing descriptions as well as illustrations of 487 species of medicinal plants (Fig. 2.1). Valerius Cordus (1515-1544), whose tragic early death prevented him from becoming the greatest of all herbalists, undertook to study plants from living material. He travelled in the

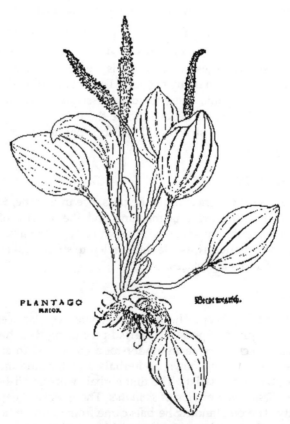

Fig. 2.1 Illustration of *Plantago major* from Fuch's *De Historia stirpium* (1542) (reproduced from Arber, *Herbals, Their Origin and Evolution*, 1938: used with permission from Cambridge University Press.)

forests of Germany and Italy, where unfortunately he fell ill and died at the young age of 29. His work *Historia plantarum*, published in 1561, many years after his death, contained accurate descriptions of 502 species, 66 apparently new. He was perhaps the first to show how to describe plants from nature accurately. Unfortunately Konrad Gesner, the editor of his work, chose to add illustrations, which were not only of poor quality, but also wrongly identified, and the work suffered for no fault of Valerius Cordus.

At the times when herbals flourished in Germany Pierandrea Mathiola was active in Italy producing *Commentarii in sex libros Pedacii Dioscorides* in 1544, adding many illustrations, though it was a commentary on Dioscorides. The **Dutch Big Three**—Rembert Dodoens, Carolus Clusius and Mathias de L'obel—spread botanical knowledge to Holland and France through their herbals. William Turner in his *Herball* (1551-1568) swept out many old superstitions concerning plants. *Herball* won for Turner the title of **Father of British Botany**.

Herbalism also saw the emergence of the **doctrine of signatures**, a result of the urge to search for clues from plants as to their nature. Many medicinal plants, the doctrine held, are stamped with a clear indication of their medicinal use. This was based upon the belief that plants and plant parts cured that portion of the human body which they resembled. Thus herbs with yellow sap would cure jaundice, the walnut kernel would comfort the brain and maidenhair fern would prevent baldness. Paracelsus and Robert Turner were the main proponents of this doctrine, later ridiculed when more knowledge concerning medicinal plants was acquired from the 17th century onwards.

Early Taxonomists

With renewed interest in plants and extensive explorations of Europe, Asia, Africa and the New World, the list of plant names increased enormously, signifying the need for a formalised scheme of classification, naming and description of plants. Botany, hitherto dependent on medicine, started to spread its wings as a science per se.

Andrea Caesalpino (1519-1603)—**the first plant taxonomist**. An Italian botanist who studied botany under Luca Ghini and became the Director of the Botanical Garden and later professor of botany and medicine at Bologna, he went to Rome in 1592 as the personal physician to Pope Clement VIII. He prepared a herbarium of 768 well-mounted plants in 1563, which is still preserved in the Museum of Natural History at Florence. His work *De Plantis libri* in 16 volumes appeared in 1583 and contained descriptions of 1520 species of plants grouped as herbs and trees and further differentiated on fruit and seed characters. Caesalpino subscribed to Aristotelian logic, taking decisions based on reasoning and not the study of features. It was not surprising, therefore, that he considered pith akin to spinal cord of animals and leaves having the sole role of protecting the apical bud. However, he highlighted the significance of reproductive characters, an attitude not liked by his contemporaries, but having much bearing on the subsequent classifications of Ray, Tournefort and Linnaeus.

Joachin Jung (1587-1657)—**The first terminologist.** A brilliant teacher in Germany, he succeeded in defining several terms such as nodes, internodes, simple and compound leaves, stamens, styles, capitulum composed of ray and disc florets. Though he left no publications of his own, two of his pupils left records of his teaching.

Gaspard (Caspar) Bauhin (1560-1624)—**Legislateur en botanique.** A Swiss botanist, he travelled extensively and formed a herbarium of 4000 specimens. He published *Phytopinax* (1596), *Prodromus theatri botanici* (1620) and lastly *Pinax theatri botanici* (1623) containing a list of 6000 species of plants giving **synonyms** (other names used for a species by earlier authors) and introducing **binomial nomenclature** for several species which he named. He sought to clarify in a single publication the confusion regarding multiplicity of names for all species known at that time. Although he did not describe genera, he did recognise difference between species and genera and several species were included under the same generic name. His elder brother Jean Bauhin (1541-1613) had earlier compiled a description of 5,000 plants with more than 3,500 figures, a work, published under the name *Historia plantarum universalis* in 1650-51, several years after his death. It is tragic that the two brothers never collaborated and worked on identical lines independently.

John Ray (1627-1705).   Ray was an English botanist who travelled extensively in Europe and published numerous works, the most significant being *Methodus plantarum nova* (1682) and *Historia plantarum* (1686-1704), a three-volume work. The last edition of *Methodus* published in 1703, included 18000 species. Ray divided the plant kingdom as shown in the outline of his classification presented in Table 2.1.

Table 2.1 Outline of classification of plants published by John Ray in *Historia plantarum* (1686-1704)

1. Herbae (Herbs)
A. Imperfectae (Cryptograms)
B. Perfectae (Seed plants)
i. Monocotyledons
ii. Dicotyledons
2. Arborae (Trees)
A. Monocotyledons
B. Dicotyledons

He was the first to group together plants that resembled one another and separated those that differed more. His classification was a great advance. It was evidently ahead of his time, groping at what later developed as natural systems, which were perfected by de Jussieu, de Candolle and Bentham and Hooker.

J.P. de Tournefort (1656-1708)—Father of genus concept. A French botanist, he studied under Pierre Magnol in the University of Montpellier and later became professor of botany at Jardin du Roy in Paris and later Director of de Jardin des Plantes in Paris. He published *Elements de botanique* in 1694, including 698 genera and 10,146 species. A Latin translation of this with additions was published as *Institutions rei herbariae* in 1700. He travelled extensively in Greece and Asia Minor and brought back 1356 plants, which were fitted into his arrangement by his admirers. He was perhaps the first to give names and description of genera, merely listing the species. Casper Bauhin who did recognise genera and species provided no such description. Tournefort was thus the first to establish genera. His system of classification, though inferior to that of Ray's, was useful for identification in recognising petaliferous and apetalous flowers, free and fused petals, and regular and irregular flowers. No doubt the system became very popular in Europe during the 18th century.

II. SEXUAL SYSTEM

A turning point in the classification approach was establishing the fact of sexuality in flowering plants by Camerarius in 1694. He concluded that stamens were male sex organs and pollen was necessary for seed set. He showed that the style and ovary form female sex organs of a flower. The thought about sexuality in plants, ridiculed by the church hitherto, once established saw renewal in botanical interest, amply exploited by Linnaeus for classifying flowering plants.

Carolus Linnaeus (Carl Linnaeus, Carl Linne, Carl Von Linne) (1707-1778)—Father of taxonomy

Whereas Darwin dominated botanical thinking during the 19th century, Linnaeus did so during the 18th. Carl Linne, Latinised as Carl Linnaeus or Carolus Linnaeus (Fig. 2.2), born in Rashult, Sweden on 23 May , 1707, had botany attached to him at birth since Linnaeus is the Latin for Linn or Linden tree (*Tilia*). His father, a country Parson, wanted his son to become a priest but Linnaeus chose to enter University of Lund in 1727 to learn Medicine. Although he had no money to buy books, his dedication impressed Professor Kilian Stobaeus, who not only allowed him full use of his library but also gave him free boarding at his house. Lund not being a suitable place for Medicine, he shifted to the University of Uppsala in 1729. In recognition of his enthusiasm for plants, Dean Olaf Celsius introduced him to a

Fig. 2.2 Carolus Linnaeus (1707-1778), the Father of taxonomy (reproduced with permission from Royal Botanic Gardens, Kew).

botanist Professor Rudbeck. Under the able guidance of Prof. Rudbeck, he published his first paper on sexuality of plants in 1729. Following favourable publicity of his paper, he was appointed as Demonstrator and subsequently promoted as Docent. In 1730, he published *Hortus upplandicus*, enumerating plants in Uppsala Botanical Garden according to Tournefort's system. Faced with problem of increasing numbers of plants which he found hard to fit in Tournefort's system, he published a revised edition of *Hortus upplandicus* with plants classified according to his own sexual system.

Linnaeus was sent on an expedition to Lapland in 1732, a trip that widened his knowledge. He brought back 537 specimens. The results of the expedition were later published as *Flora lapponica* (1737).

Linnaeus went to the Netherlands in 1735 and obtained an M.D. degree from the University of Haderwijk. While in the Netherlands, he met several prominent naturalists including John Frederick Gronovius and Hermann Boerhaave, the former financing publication of *Systema naturae* (1735) presenting an outline of the **sexual system** of Linnaeus. He became the personal physician of a wealthy person George Clifford, Director of the Dutch East India Company, and this gave Linnaeus an opportunity to study numerous tropical and temperate plants grown by Clifford in his garden. It was at Clifford's expense that Linnaeus published several manuscripts, including *Hortus cliffortianus* and *Genera plantarum* (1737). He went to England, where he met Professor John Jacob Dillen, who initially thought of Linnaeus as 'this is he who is bringing all botany into confusion', but he soon became the advocate of the Linnaean system in England. He also met the de Jussieu brothers in France.

Following the death of Professor Rudbeck, Linnaeus was appointed Professor of medicine and botany at the University of Uppsala, a position he held until his death in 1778. He published his best known *Species plantarum* in 1753. His growing fame and publications attracted a larger number of students, their number increasing every year and the botanical garden at Uppsala was enriched considerably.

His botanical excursions every summer also included an annotator to take notes, a Fiscal to maintain discipline and marksmen to shoot birds. At the end of each trip, they marched back to the town with Linnaeus at the head, with French horns, kettledrums and banners.

In recognition of his contributions, Linnaeus was made **Knight of the Polar Star** in 1753, the first Swedish scientist to get this honour. In 1761, he was granted the **patent of nobility** and from this date came to be known as **Carl von Linne**.

Among his enthusiastic students were Peter Kalm and Peter Thunberg. Kalm collected extensively in Finland, Russia and America and when he returned with bundles of collection from America, Linnaeus was bedridden, but forgot his ailment and transferred his concern to plants. Thunberg collected extensively in Japan and South Africa.

Linnaeus first outlined his system in *Systema naturae*, which classified all known plants, animals and minerals. In his *Genera plantarum* he listed and described all plant genera known to him. In *Species plantarum* he listed and described all species of plants. For each species there was (Fig. 2.3):

  (i) a generic name;
 (ii) a polynomial descriptive phrase or phrase-name commencing with generic name and of up to twelve words, intended to serve as description of the species;
(iii) a trivial name or specific epithet on the margin;
 (iv) synonyms with reference to important earlier literature;
  (v) habitats and countries.

The generic name followed by trivial name formed the name for each species. Linnaeus thus established the **binomial nomenclature**, first started by Casper Bauhin and the generic concept, started by Tournefort.

The system of Linnaeus, very simple in application, recognised 24 classes (Table 2.2), mostly on the basis of stamens. These classes were further subdivided based on carpel characteristics into orders such as Monogynia, Digynia, etc. Such a classification based on stamens and carpels resulted in artificial grouping of unrelated taxa and separation of relatives.

Linnaeus knew that his system was more convenient than natural, but it was the need of the day when there was a tremendous increase in the number of plants known to man, and there was a need for quick identification and placement. This is exactly what the **sexual system** of Linnaeus

DIADELPHIA HEXANDRIA. 699
Classis XVII

DIADELPHIA.

HEXANDRIA,
FUMARIA.
*Corollis bicalcaratis,

I.   FUMARIA Scapo nudo. *Hort. Cliff.* 251:   **Gron.*   *cucullaria,*
     virg. 171. Rov.lugdb,. 393
     Fumaria tuberosa insipida. *Corn. canad. 127.*
     Fumaria siliquosa, radice grumosa, flore bicorporeo ad
     labia conJucto, virginiana. *Plak, alam. I62. t. 90. f*
     3· *Raj. suppl.*   475·
     Cucullaria. *Juss. act. paris* . 1743·
     Habitat in Virginia, Canada   4
     Radix *tuberosa;* Folium *radicale tricompositum.* Scapus
     nudus, Racemo simplici; bracteae vix ullae; Nectarium
     duplex corollam basi bicornem efficiens.

2.   FUMARIA floribus *postice* bilobis, caule folioso.   *spectabilis.*
     Habitat in Sibiria. *D. Demidoff:*
     Planta *eximia floribus speciofissimis, maximis.* Habitus
       Fumariae bulbosae, sed majora omnia. Rami *ex alis rarioris.*
       Caulis *erectus.* Racemus *absque bracteis.*
       Corollae *magnitudine extimi articuli pollicis, pone* in *duos
       lobos aequales, rotundatos divisae.*

*Corollis unicalcaratis.

3.   FUMARIA caule simplici, bracteis longitudine florum   *bulbosa.*

Fig. 2.3 A portion of a page from *Species plantarum* of Linnaeus (1753). Specific
epithet (trivial name) is indicated towards the margin.

achieved with merit. His *Species plantarum* (1753) marks the starting point of
botanical nomenclature today. Linnaeus did aim at natural classification
and in the 6th edition of his *Genera plantarum* (1764) he appended a list of 58
natural orders. It was, however, left to others to carry forward. Linnaeus
had done his job according to the demands of the day.

Following his death in 1778, his son Carl got the post of Professor as well
as the collections at Uppsala. When he died in 1783, the collections went to
the widow of Linnaeus, whose sole aim was to sell it to the highest bidder.
Fortunately this highest bidder of 1000 guineas was J.E. Smith, an English
botanist. Smith founded the Linnaean Society of London in 1788 and hand-
ed over the herbarium to this society. Herbarium specimens have since been
photographed and are available in microfiche.

The Linnaean classification remained dominant for a long time. The 5th
edition of *Species plantarum* appeared as late as in 1797-1805, greatly enlarged and edited by C.L. Wildenow in four large volumes.

Table 2.2 Outline of the 24 classes recognised by Linnaeus in his *Species plantarum* (1753) on the basis of stamens

Class 1. Monandria—stamen one
2. Diandria—stamens two
3. Triandria—stamens three
4. Tetrandria—stamens four
5. Pentandria—stamens five
6. Hexandria—stamens six
7. Heptandria—stamens seven
8. Octandria—stamens eight
9. Ennandria—stamens nine
10. Decandria—stamens ten
11. Dodecandria—stamens 11-19
12. Icosandria—stamens 20 or more, on the calyx
13. Polyandria—stamens 20 or more, on the receptacle
14. Didynamia—stamens didynamous; 2 short, 2 long
15. Tetradynamia—stamens tetradynamous; 4 long, 2 short
16. Monadelphia—stamens monadelphous; united in 1 group
17. Diadelphia—stamens diadelphous; united in 2 groups
18. Polyadelphia—stamens polyadelphous; united in 3 groups
19. Syngelnesia—stamens syngenesious; united by anthers only
20. Gynandria—stamens united with the gynoecium
21. Monoecia—plants monoecious
22. Dioecia—plants dioecious
23. Polygamia—plants polygamous
24. Cryptogamia—flowerless plants

III. NATURAL SYSTEMS

Linnaeus had provided a readily referable cataloguing scheme for a large number of plants, but it soon became evident that unrelated plants came together in such groupings. A need was realised for a more objective classification. France, which was undergoing an intellectual ferment and where the Linnaean system never became popular, took the lead in developing natural systems of classification.

Michel Adanson (1727-1806)

A French botanist, unimpressed with artificial choice of characters, he devised a classification of both animals and plants, on equal use of as many features as possible. In his two-volume work *Familles des plantes* (1763), he recognised 58 **Natural orders** according to their natural affinities. Present-day **Numerical taxonomy** is based on the idea conceived by Adanson and now developed into **Neo-Adansonian principles**.

Jean B.P. Lamarck (1744-1829)

A French naturalist, he authored *Flore Française* (1778), which in addition to a key for identification of plants, contained principles concerning natural grouping of species, orders and families. He is better known for his evolutionary theory, **Lamarckism**.

de Jussieu family

Four well-known botanists belonged to this prominent French family. Of the three brothers—Antoine (1686-1758), Bernard (1699-1776) and Joseph (1704-1779)—the youngest spent many years in South America, where after losing his collections of five years, he became insane. The elder two studied at the University of Montpellier under Pierre Magnol. Antoine succeeded Tournefort as Director **de Jardin des Plantes, Paris** and later added Bernard to the staff.

Bernard started arranging plants in the garden according to the Linnaean system, introducing changes, so that when finally set, it had no resemblance with the Linnaean system. He never published his system and it was left to his nephew Antoine Laurent de Jussieu (1748-1836) (Fig. 2.4) to publish this classification along with his own changes in *Genera plantarum* (1789).

Fig. 2.4   Antoine Laurent de Jussieu (1748-1836) the author of *Genera plantarum* (1789) largely based on the work of his Uncle Bernard de Jussieu (reproduced with permission from Royal Botanic Gardens, Kew).

An outline of the classification is presented below:
1. Acotyledones
2. Monocotyledones

3. Dicotyledones
    i.   Apetalae
    ii. Monopetalae
    iii. Polypetalae
    iv. Diclines irregulares

In this classification the plants were divided into three groups, further divided on corolla characteristics and ovary position to form 15 classes and 100 orders (until the beginning of the present century, class and order were mostly used as names of categories now understood as order and family respectively). **Acotyledones** in addition to cryptograms contained some hydrophytes whose reproduction was not known then. **Diclines irregulares** contained Amentiferae, Nettles, Euphorbias as also the Gymnosperms.

de Candolle family

The de Candolles were a Swiss family of botanists. Augustin Pyramus de Candolle (1778-1841) was born in Geneva, Switzerland but took his education in Paris, where he became the Professor of Botany at Montpellier (Fig. 2.5). He published several books, the important being *Theorie elementaire de la botanique* (1813), wherein he proposed a new classification scheme, outlined the important principles and introduced the term **taxonomy**. From 1816 until his death he worked in Geneva and undertook a monumental work, intended to describe every known species of vascular plants under the title *Prodromus systematis naturalis regni vegetabilis*, the first volume appearing in 1824. He published seven volumes himself. His son Alphonse de Candolle and grandson Casimir de Candolle continued the work. Alphonse

Fig. 2.5 Augustin Pyramus de Candolle (1778-1841) who first introduced the term 'taxonomy' in his *Theorie elementaire de la botanique* (1813) (reproduced with permission from Royal Botanic Gardens, Kew).

published ten more volumes, the last one in 1873, resulting in revision of several families by specialists.

The classification by A.P. de Candolle delimited 161 natural orders (the number was increased to 213 in the last revision of *Theorie elementaire....*, edited by Alphonse in 1844), grouped primarily on the basis of the presence or absence of vascular structures (Table 2.3).

Table 2.3 Outline of classification proposed by A.P. de Candolle in his *Theorie elementaire de la botanique* (1813)

```
  1. Vasculares (vascular bundles present)
        Class 1. Exogenae (dicots)
               A.  Diplochlamydeae
                      Thalamiflorae
                      Calyciflorae
                      Corolliflorae
               B.  Monochlamydeae (also including Gymnosperms)
        Class 2. Endogenae
               A.  Phanerogamae (monocots)
               B.  Cryptogamae
 II. Cellulares (no vascular bundles)
        Class 1. Foliaceae (Mosses, Liverworts)
        Class 2. Aphyllae (Algae, Fungi, Lichens)
```

Ferns were provided a place co-ordinate with monocots and in contrary to de Jussieu, Gymnosperms were given a place, although among dicots. The importance of anatomical features was highlighted and successfully employed in classification.

Robert Brown (1773-1858)

An English botanist, who did not propose a classification of his own but demonstrated that Gymnosperms were a group discrete from dicotyledons and had naked ovules. He also clarified the floral morphology and pollination of Asclepiadaceae and Orchidaceae.

George Bentham (1800-1884) and Sir J.D. Hooker (1817-1911)

These two English botanists jointly produced the most elaborate natural system of classification in their three-volume work *Genera plantarum* (1862-83), a classification of seed plants recognising Dicotyledons, Gymnosperms and Monocotyledons as distinct groups. In all 202 natural orders (now treated as families) were recognised grouped into cohorts (now orders) and series. The descriptions were based on original studies of specimens and dissections of plants themselves, and did not represent a mere compilation of existing literature.

Genera plantarum, though published after the *Origin of species* by Darwin, is pre-Darwinian in concept. The species were regarded as fixed entities, not changed through time. The descriptions were accurate, authentic, and no doubt the system is still useful and followed in several major herbaria of the world.

IV. PHYLOGENETIC SYSTEMS

The publication of *The Origin of Species* by Charles Darwin, with every copy of the first edition sold on the first day, 24 November 1859, revolutionised biological thinking. The species was no longer regarded as a fixed entity having remained unchanged since its creation. Species were now looked upon as systems of populations, which are dynamic and change with time to give rise to lineages of closely related organisms. Once the existence of this evolutionary process was acknowledged, the systems of de Candolle as also of Bentham and Hooker were found inadequate and classifications, which made an attempt to reconstruct evolutionary sequence, found immediate takers.

A. Transitional systems

The early systems were not intended to be phylogenetic. They were attempts to rearrange earlier natural systems in the light of prevalent phylogenetic theories.

A.W. Eichler (1839-1887).   A German botanist who proposed the rudiments of a system in 1875. This was elaborated into a unified system covering the entire plant kingdom and finally published in the third edition of *Syllabus der vorlesungen...* (1883). The plant kingdom was divided into two subgroups: Cryptogamae and Phanerogamae, the latter further subdivided into Gymnospermae and Angiospermae. Angiospermae was divided into two classes: Monocotyledons and Dicotyledons. Only two groups **Choripetalae** and **Sympetalae** were recognised in Dicotyledons. Gymnosperms thus found their separate identity before angiosperms, Monochlamydeae found itself abolished and dispersed among the two groups. Monocotyledons, strangely, found a place before Dicotyledons.

Adolph Engler (1844-1930) and **Karl Prantl** (1849-1893).   These two German botanists developed a system of classification based on Eichler and differing in details. The monumental work *Die naturlichen pflanzenfamilien* (1887-1915) in 23 volumes covered the whole plant kingdom. The work included the keys and descriptions of all known genera of plants. The seed plants (Embryophyta siphonogama) were divided into Gymnospermae and Angiospermae, and the latter into Monocotyledoneae and Dicotyledoneae. The two subclasses of dicots were named as **Archichlamydeae** (petals absent or free) and **Metachlamydeae** (petals fused).

The Englerian system with its improvements over Bentham and Hooker, soon replaced it in several European herbaria. Eichler and Engler had, however, misread the evolutionary sequence, as was evidenced by subsequent palaeobotanical and anatomical evidence. **Amentiferae**, considered by Engler to be the most primitive group in dicots owe their simplicity to evolutionary reduction and not primitiveness. Monocots are advanced and not primitive in comparison to dicots. Dichlamydeous flowers (with sepals and petals) are now considered primitive and not advanced over monochlamydeous flowers (perianth only).

Although the systems of Bentham and Hooker as also of Engler and Prantl are now obsolete in the light of current phylogenetic views, they continue to be the most dominant systems followed in Floras and Herbaria, on the strength of their thorough treatment, enabling one to identify and assign any genus to its family. Unfortunately no subsequent publication on classification of flowering plants has been that thorough, and although we may have very sound systems of classification, our Floras and Herbaria will continue to follow one of these two systems only for years to come.

B. Intentional phylogenetic systems

The natural systems rearranged in the light of phylogenetic information soon gave way to systems that reflect evolutionary development. A beginning in this direction was made by an American botanist, Charles Bessey.

Charles Bessey (1845-1915). The foundations of modern phylogenetic classifications were laid by C.E. Bessey (Fig. 2.6), a student of Asa Gray and later Professor of botany at the University of Nebraska. He was the first American to make a major contribution to plant classification, and the first botanist to develop intentional phylogenetic classification. He based his classification on Bentham and Hooker, modified in the light of his 28 **dicta** and published in *Ann. Mo. Bot. Gard.* under the title 'Phylogenetic Taxonomy of flowering plants' (1915). He considered angiosperms to have evolved monophyletically from Cycadophyta belonging to implied **bennettitalean ancestry** (Table 2.4).

Table 2.4 Outline of the classification of angiosperms proposed by Charles Bessey (1915).

Class 1. Alternifoliae (Monocotyledoneae)
    Subclass 1. Strobiloideae (5 orders)
    Subclass 2. Cotyloideae (3 orders)
Class 2. Oppositifoliae (Dicotyledoneae)
    Subclass 1. Strobiloideae
        Superorder 1. Apopetalae-polycarpellatae (7 orders)
        Superorder 2. Sympetalae-polycarpellatae (3 orders)
        Superorder 3. Sympetalae-dicarpellatae (4 orders)
    Subclass 2. Cotyloideae
        Superorder 1. Apopetalae (7 orders)
        Superorder 2. Sympetalae (3 orders)

Fig. 2.6 Charles Bessey (1845-1915) who initiated the modern phylogenetic systems of classification. He proposed his ideas in *Ann. Mo. Bot. Gard.* (1915) (reproduced with permission from Royal Botanic Gardens, Kew).

Bessey believed in the **strobiloid theory** of origin of the flower, the latter having originated from a vegetative shoot with spiral phyllomes, of which some modified to form sterile perianth, fertile stamens and carpels. Two evolutionary lines from such a flower formed **Strobiloideae** (Ranalian line) with connation of like parts and **Cotyloideae** (Rosalian line) with connation of unlike parts. Ranales in dicots and Alismatales in monocots were considered the most primitive in each group, a fact recognised by most subsequent authors. Ranalian plants were considered primitive angiosperms having given rise to monocots, but unfortunately monocots were placed before dicots.

Bessey also initiated representing evolutionary relationships through an evolutionary tree with primitive groups at the base and the most advanced at the tips of branches (Fig. 2.7). His diagram resembling a cactus plant is better known as the **Besseyan cactus**.

Hans Hallier (1868-1932).
Hallier was a German botanist who developed a classification resembling Bessey's and starting with Ranales. Dicots were, however, placed before monocots. **Magnoliaceae** were separated from Ranales and placed in a separate order, **Annonales**.

Wettstein (1862-1931).
An Austrian systematist who published his classification in *Handbuch der systematischen botanik* (1930, 1935). The classification resembled that of Engler's in considering unisexual flowers primitive but treated monocots advanced over dicots, considered **Helobiae** primitive and Pandanales advanced. Many of his conclusions on phylogeny have been adopted in subsequent classifications.

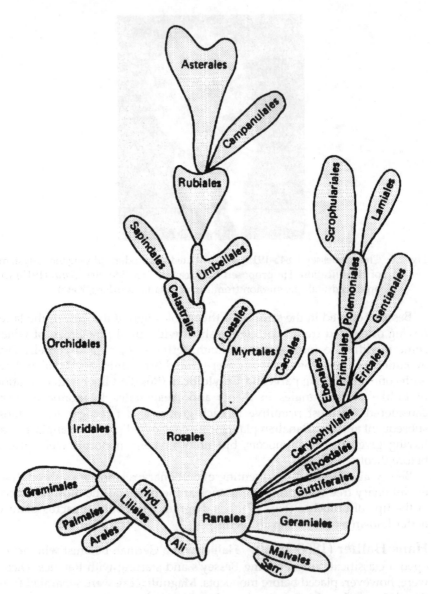

Fig. 2.7 **Besseyan cactus** or **Opuntia Besseyi** showing relationship of orders
recognised by Bessey.

Alfred Rendle (1865-1938).   An English botanist associated with the
British Museum of Natural History, he published *Classification of Flowering
Plants* (1904, 1925), resembling that of Engler in considering monocots more
primitive than dicots and Amentiferae a primitive group under dicots. He

recognised three **grades** in dicots: **Monochlamydeae, Dialapetalae** (petals free) and **Sympetalae**. In monocots Palmae were separated as a distinct order and Lemnaceae considered advanced over Araceae.

John Hutchinson (1884-1972). A British botanist associated with the Royal Botanic Gardens at Kew, he developed a system of classification which appeared in its final form in 3rd edition of his *The Families of Flowering Plants* (1973). The classification was based on 24 **principles**, recognising 420 families of flowering plants. Dicots are considered more primitive than monocots, and divided into two groups: **Lignosae** and **Herbaceae**. Magnoliales are considered to be the most primitive dicots. The division of dicots into woody and herbaceous lines has, however, received wide criticism. Several closely related families have been separated and placed in distant groups.

Lyman Benson. He developed a classification designed for teaching and published in his book *Plant Classification* (1957). Dicotyledons are divided into five groups distinguished as outlined below:
1. Thalamiflorae (hypogynous, free or no petals)
2. Corolliflorae (hypogynous, petals fused)
3. Calyciflorae (perigynous or epigynous, petals free or none)
4. Ovariflorae (epigynous, petals fused)
5. Amentiferae (catkin bearing)

Monocotyledons are divided directly into 13 orders, starting with Alismales and ending with Pandanales. Although several realignments have been made by Benson, de Candolle as also Bentham and Hooker have been followed for grouping in dicots and Bessey's outline for classification of monocots.

C. Contemporary phylogenetic systems

A number of contemporary workers are involved in improving schemes of classification based on new information from various sources. Recent data from palaeobotany, phytochemistry, ultrastructure and improved techniques of numerical analysis of available data have helped in developing classifications that have several features in common, though differing in some basic concepts. It is now largely agreed that Angiosperms are a **monophyletic** group with dicots being more primitive than monocots. Vesselless **Winteraceae** are now generally regarded as most primitive among living angiosperms.

Armen Takhtajan (1910-1997). A Russian botanist who developed a classification system of flowering plants, which were periodically revised, the last version being published in 1997. He placed angiosperms in Division· **Magnoliophyta** divided into two classes **Magnoliopsida** (dicots) and **Liliopsida** (monocots), divided in turn into subclasses. He first proposed his classification in 1954, then followed by major revisions in 1966, 1980,

1987 and in 1997 finally recognised 11 subclasses in dicots and 6 in monocots. These are further divided into superorders (ending in -anae), orders and families; 56 superorders, 175 orders and 459 families are included in dicots, whereas monocots include 16 superorders, 58 orders and 133 families. Phylograms (unfortunately missing in 1997 version) are provided indicating the putative relationship of groups, latter being represented by balloons or bubbles (hence the name 'bubble diagram') of different sizes indicating the relative number of species in the groups. Because of the elaborate bubble diagram and its unique display Woodland (1991) has aptly named it 'Takhtajan's flower garden'.

Arthur Cronquist (1919-1992).   While with the New York Botanical Garden, Cronquist first proposed his classification in 1957 and published major revisions in 1968, 1981,and 1988; he recognised 6 subclasses in dicots and 5 in monocots. His classification closely resesembles Takhtajan in placing angiosperms in Division **Magnoliophyta** divided similarly into **Magnoliopsida** (dicots) and **Liliopsida** (monocots). Superorders, however, are not recognised. Dicots are divided into 64 orders and 318 families, monocots into 19 orders and 65 families. Whereas Takhtajan attaches more importance to cladistics, Cronquist attaches more weightage to phenetic relationship in the placement of various groups. The **phylogram** in the form of a bubble diagram has also been proposed but not as elaborate as that of Takhtajan, since it depicts only the major groups of angiosperms.

C.R. de Soo.   From Budapest, Hungary, de Soo proposed (1975) a classi-fication essentially similar to Takhtajan but preferring the name **Angiospermophyta** for angiosperms, **Dicotyledonopsida** for dicots and **Monocotyledonopsida** for monocots. Five subclasses are included in dicots and 3 in monocots; 54 orders in dicots and 14 in monocots are recognised.

Robert Thorne (1920-) (Fig. 2.8)   An American taxonomist who has developed and periodically revised a system of classification that approaches the above two systems but differs in that angiosperms are ranked as a class and dicots and monocots as subclasses, which in turn are divided directly into superorders. Thorne first put forward his classfication in 1968 and proposed revisions in 1976, 1981, 1983 and 1992. He earlier preferred the ending -florae over -anae of Takhtajan for superorders, but has now (1992 onwards) accepted the ending 'anae'. He incorporated the role of phytochemistry in realignment of taxa, recognising subfamilies more frequently and applied the principle of priority up to the class rank, thus preferring the name **Annonopsida** for angiosperms, **Annonidae** for dicots, replacing Magnoliflorae by **Annoniflorae** and Magnoliales by **Annonales**. Since 1992 he has, however, abandoned this departure from contemporary systems and adopted the generally accepted names **Magnoliopsida**, **Magnoliidae** and **Magnoliales**. Nineteen superorders are recognised in dicots and nine in monocots.

Fig. 2.8   Robert Thorne of Rancho Santa Botanic Garden.

The Thorne diagram indicating the relationship between different groups is a **phylogenetic shrub** viewed from above, with the centre of the diagram left empty to indicate extinct early angiosperms, those nearer the centre being the primitive groups and those nearer the periphery the advanced ones. The relative number of species in different groups is indicated by balloons of different sizes.

Rolf Dahlgren (1932-1987) (Fig. 2.9).   A Danish botanist who first proposed his classification in 1975 with revisions in 1980, 1981, 1983. The classification is closely similar to Thorne in using the name **Magnoliopsida** for angiosperms, **Magnoliidae** for dicots, and Liliidae for monocots. An updated revision of his classification was published by his wife Gertrud Dahlgren (1989). The realignments are based on a large number of phenetic characteristics, mainly phytochemistry, ultrastructure and embryology. The system includes 25 superorders in dicots and 8 in monocots. Dahlgren's diagram is similarly a cross-section of the **phylogenetic tree** from near the top, a diagram that is excellent for mapping of character distribution in various groups which may go a long way in developing an ideal classification system. Already several hundred such maps have been developed by Dahlgren and his associates. Dahlgren pointed out that recognition of Dicotyledons and Monocotyledons would not be allowed if one followed rigid **cladistic** approaches, but he nevertheless considered Monocotyledons a unique group worthy of subclass rank. This classification also initially preferred the ending **-florae** for superorders but Gertrud Dahlgren (1989) shifted to the usage of **-anae** since the ending **-florae** restricts usage to angiosperms and is not universal in application.

Fig. 2.9 Rolf Dahlgren and his wife Gertrud Dahlgren who has continued his work on the classification of Angiosperms since his death in 1987. Gertrud has concentrated on evolutionary botany and species differentiation after 1990 (photographs courtesy Gertrud Dahlgren).

Bremer and Wanntorp (1981) did propose a **cladistic classification** suggesting that angiosperms should be treated as subclass **Magnoliidae** of class **Pinatae** (seed plants). They argue that Monocotyledons and Dicotyledons should not be recognised because it will make the group **paraphyletic**, suggesting that angiosperms should be directly divided into a number of superorders.

Mention should also be made of the classification schemes of Sporne (1976) and Stebbins (1974). Although largely based on the Takhtajan-Cronquist pattern, they present the relationship between the groups in the form of a **circular diagram**, the degree of specialisation of a group indicated by an **advancement index**, zero at the centre and one hundred along the periphery. As expected, the centre of such a diagram is empty, since the earliest angiosperms are now extinct.

3

Botanical Nomenclature

Nomenclature deals with the application of a correct name to a plant or a taxonomic group. In practice it is often combined with identification, since while identifying an unknown plant specimen the author chooses and applies the correct name. The favourite temperate plant is correctly identified whether you call it 'Seb' (vernacular Hindi name), Apple, *Pyrus malus* or *Malus malus*, but only by using the correct scientific name *Malus pumila* does one combine identification with nomenclature. The current activity of botanical nomenclature is governed by the International Code of Botanical Nomenclature (**ICBN**) published by the International Association of Plant Taxonomy (**IAPT**). The Code is revised after changes at each International Botanical Congress. The naming of the animals is governed by the International Code of Zoological Nomenclature (**ICZN**) and that of bacteria by the International Code for the Nomenclature of bacteria (**ICNB**). Naming of cultivated plants is governed by the International Code of Nomenclature for Cultivated Plants (**ICNCP**), which is largely based on **ICBN** with a few additional provisions.

I. NEED FOR SCIENTIFIC NAMES

Scientific names formulated in Latin are preferred over vernacular or common names since the latter pose a number of problems:

1. Vernacular names are not available for all the species known to man.
2. Vernacular names are restricted in usage and are applicable in one or a few languages only. They are not universal in application.
3. Common names usually do not provide information indicating family or generic relationship. Roses belong to the genus *Rosa*, woodrose is a member of the genus *Ipomoea* and primrose belongs to the genus *Primula*. The three genera in turn belong to three different families—Rosaceae, Convolvulaceae and Primulaceae, respectively.

4. Frequently, especially in widely distributed plants, many common names may exist for the same species in the same language in the same or different localities. Cornflower, blue bottle, bachelor's button and ragged robin all refer to the same species *Centaurea cyanus*.
5. Often two or more unrelated species are known by the same common name. Bachelor's button may thus be *Tanacetum vulgare, Knautia arvensis* or *Centaurea cyanus*. Cockscomb is similarly a common name for *Celosia cristata* but is also applied to a seaweed *Plocamium coccinium* or to *Rhinanthus minor*.

II. HISTORY OF BOTANICAL NOMENCLATURE

For several centuries the names of plants appeared as polynomials—long descriptive phrases, often difficult to remember. A species of willow, for example, was named *Salix pumila angustifolia altera* by Clusius in his herbal (1583). Casper Bauhin (1623) introduced the concept of **Binomial nomenclature** under which the name of a species consists of two parts, first the name of the genus to which it belongs and the second the **specific epithet**. Onion is thus appropriately named *Allium cepa, Allium* being the generic name and *cepa* the specific epithet. Bauhin, however, did not use binomial nomenclature for all the species and it was left to Carolus Linnaeus to establish this system of naming firmly in his *Species plantarum* (1753). The early rules of nomenclature were set forth by Linnaeus in his *Critica botanica* (1737) and further amplified in *Philosophica botanica* (1751). A.P. de Candolle in his *Theorie elementaire de la botanique* (1813) gave explicit instructions on nomenclatural procedures, many taken from Linnaeus. Steudel in *Nomenclator botanique* (1821) provided Latin names for all flowering plants known to the author together with their synonyms.

The first organised effort was made by Alphonse de Candolle who circulated a copy of his manuscript *Lois de la nomenclature botanique* and after deliberations of the First International Botanical Congress at Paris (1867), the **Paris Code**, also known as 'de Candolle rules' was adopted. Linnaeus (1753) was made the starting point for plant nomenclature and rule of priority was made fundamental. Not satisfied with the Paris Code, American botanists adopted a separate **Rochester Code** (1892), which introduced the concept of **types**, strict application of rules of priority even if the name was a **tautonym** (specific epithet repeating the generic name, e.g. *Malus malus*).

The Paris Code was replaced by the **Vienna Code** (1905) which established *Species plantarum* (1753) of Linnaeus as the **starting point**, tautonym was not accepted and **Latin diagnosis** was made essential for new species. In addition, a list of conserved generic names (**Nomina generica conservanda**) was approved. Not satisfied with the Vienna Code also, adherents of the Rochester Code adopted the **American Code** (1907) which did not accept the list of conserved names and the requirement for Latin diagnosis.

It was not until the 5th International Botanical Congress at **Cambridge** (1930) that the differences were finally resolved and a truly International Code evolved, accepting the concept of type method, rejecting the tautonyms, making Latin diagnosis mandatory for new groups and approving conserved generic names. The Code has since been constantly emended at each International Botanical Congress. The 15th International Botanical Congress was held at Tokyo in 1993. The code approved at Tokyo was published in 1994 (Greuter et al., 1994).

Publication of the Code is based on the realisation that botany requires a precise and simple system of nomenclature used by botanists in all countries. The Code aims at provision of a stable method of naming taxonomic groups, avoiding and rejecting the use of names which may cause error or ambiguity or throw science into confusion. The Code is divided into 3 divisions:

I. Principles
II. Rules and recommendations
III. Provisions for modification of the Code

In addition, the Code includes the following appendices:

I. Names of hybrids
IIA. Nomina familiarum algarum, fungorum et pteridophytorum conservanda et rejicienda
IIB. Nomina familiarum bryophytorum et spermatophytorum conservanda
IIIA. Nomina generica conservanda et rejicienda
IIIB. Nomina specifica conservanda et rejicienda
IV. Nomina utique rejicienda (B. Fungi, C. Bryophyta, E. Spermatophyta)
V. Opera utique oppressa

The last three useful appendices were included for the first time in the Tokyo Code. The first (IIIB) includes the names of conserved and rejected specific names; the second (IV) the names and all combinations based on these names, which are ruled as rejected under Art. 56, and none is to be used; and the last (V) the list of publications (and the category of taxa therein) which are not validly published according to the Code.

Principles form the basis of the system of botanical nomenclature. There are 76 main **rules** (set out as articles) and associated **recommendations**. The object of the rules is to put the nomenclature of the past into order and provide for that of the future; names contrary to the rules cannot be maintained. Recommendations deal with subsidiary points, and are meant for uniformity and clarity. Names contrary to the recommendations cannot, on that account, be rejected, but they are not examples to be followed. **Conserved names** include those that do not satisfy the principle of priority but are sanctioned for use. The various rules and recommendations are discussed here under relevant headings.

III. PRINCIPLES OF ICBN

The International Code of Botanical Nomenclature is based on the following set of six principles, which are the philosophical basis of the Code and provide guidelines for the taxonomists who propose amendments or deliberate on the suggestions for modification of the Code:

1. Botanical Nomenclature is independent of Zoological Nomenclature. The Code applies equally to the names of taxonomic groups treated as plants whether or not these groups were originally so treated (for the purpose of this Code, 'plants' does not include bacteria but does include fungi).
2. The application of names of taxonomic groups is determined by means of nomenclatural types.
3. Nomenclature of a taxonomic group is based upon priority of publication.
4. Each taxonomic group with a particular circumscription, position and rank can bear only one correct name, the earliest that is in accordance with the rules.
5. Scientific names of taxonomic groups are treated as Latin, regardless of derivation.
6. The rules of nomenclature are retroactive, unless expressly limited.

IV. NAMES OF TAXA

Taxon (pl. taxa) refers to a taxonomic group belonging to any rank. The system of nomenclature provides a hierarchical arrangement of ranks. Every plant is treated as belonging to a number of taxa, each assigned a particular rank. Onion thus belongs to *Allium cepa* (species rank), *Allium* (genus rank), Liliaceae (family rank) and so on. The ending of the name indicates its rank: ending **-phyta** denotes a division, **-phytina** a subdivision, **-opsida** a class, **-opsidae** or **-idae** a subclass, **-ales** an order, **-ineae** a suborder and **-aceae** a family. The detailed hierarchy of ranks and endings with examples is given in Table 3.1.

The endings for rank, subclass and above are recommendations whereas for order and below these are mandatory rules. The name of a family ends in **-aceae**, but there are eight families of angiosperms whose original names, not in accordance with the rules, have been sanctioned because of old traditional usage. A list of these families together with alternate names (in accordance with the rules and which need to be encouraged) is given below:

Traditional name	*Alternate name*
Cruciferae	Brassicaceae
Guttiferae	Clusiaceae
Leguminosae	Fabaceae
	(Contd.)

(contd.)

Traditional name	Alternate name
Umbelliferae	Apiaceae
Compositae	Asteraceae
Labiatae	Lamiaceae
Palmae	Arecaceae
Gramineae	Poaceae

Under a unique exception to article 52 of the Code the name Leguminosae is sanctioned only as long as it includes all the three subfamilies: Papilionoideae, Caesalpinioideae and Mimosoideae. In case these are upgraded as families, then the name Papilionaceae is conserved against Leguminosae for the first of these. The two alternate names allowed then are Papilionaceae and Fabaceae.

Table 3.1 Ranks and endings provided by ICBN

Rank	Ending	Example
Division	-phyta	Magnoliophyta
	-mycota (Fungi)	Eumycota
Subdivision	-phytina	Pterophytina
	-mycotina (Fungi)	Eumycotina
Class	-opsida	Magnoliopsida
	-phyceae (Algae)	Chlorophyceae
	-mycetes (Fungi)	Basidiomycetes
Subclass	-opsidae	Pteropsidae
	-idae (Seed plants)	Rosidae
	-physidae (Algae)	Cyanophysidae
	-mycetidae (Fungi)	Basidiomycetidae
Order	-ales	Rosales
Suborder	-ineae	Rosineae
Family	-aceae	Rosaceae
Subfamily	-oideae	Rosoideae
Tribe	-eae	Roseae
Subtribe	-inae	Rosinae
Genus	-us, -um, -is, -a, -on	*Pyrus, Allium, Arabis, Rosa, Polypogon*
Subgenus		*Cuscuta* subgenus *Eucuscuta*
Section		*Scrophularia* section *Anastomosanthes*
Subsection		*Scrophularia* subsection *Vernales*
Series		*Scrophularia* series *Lateriflorae*
Species		*Rosa canina*
Subspecies		*Crepis sancta* subsp. *bifida*
Varietas		*Lantana camara* var. *varia*
Forma		*Tectona grandis* f. *punctata*

The names of the groups belonging to ranks above the level of genus are uninomials in the plural case.

Genus

The generic name is a uninomial **singular** word treated as a noun. The genus may have a masculine, neuter or feminine form as indicated by the ending: **-us, -pogon** commonly stand for masculine genera, **-um** for neuter and **-a, -is** for feminine genera. The first letter of the generic name is always capitalised. The name may be based on any source, but the common sources for generic names are as under:

1. **Commemoration of a person** such as *Linnaea* for Linnaeus, *Bauhinia* for Bauhin, *Victoria* for Queen Victoria of England, *Zinobia* for Queen Zenobia of Palmyra and *Moltkia* for Count Moltke of Denmark. The names commemorating a person, man or woman always take the feminine form.
2. **Based on a place** such as *Araucaria* after Arauco a province of Chile, *Arabis* for Arabia and *Sibiraea* for Siberia.
3. **Based on an important character** such as yellow wood in *Zanthoxylon*, liver-like leaves in *Hepatica*, marshy habit of *Hygrophila* and spiny fruit of *Acanthospermum*.
4. **Aboriginal names** taken directly from a language other than Latin without alteration of ending. *Narcissus* is the Greek name for daffodils, *Ginkgo* a Chinese, *Vanda* a Sanskrit and *Sasa* a Japanese aboriginal name.

The generic name of a tree, whatever be the ending, takes a feminine form since trees are generally feminine in classical Latin. *Pinus, Quercus* and *Prunus* are thus all feminine genera. If two words are used to form a generic name, these have to be joined by a hyphen.

Species

The name of a species is a **binomial:** consisting of two words, a generic name followed by a specific epithet. The Code recommends that all specific epithets should begin with a lower case initial letter. An upper case initial letter is sometimes used, however, for specific epithets derived from a person's name, former generic name or a common name. The Code discourages such usage for specific epithets. A specific epithet may be derived from any source or composed arbitrarily. The following sources are commonly used.

1. **Name of a person**. The specific epithet named after a person may take genitive (possessive) or adjectival form:
   (i) When used in the genitive form the epithet takes its form depending on the ending of the person's name. For names ending in a vowel or **-er** the letter *-i* is added but if it ends in a consonant *-ii* is added, and if the name ends in **-a** then *-e* is added to the name of the person. For names of the persons already in Latin (e.g. Linnaeus) the Latin ending (-us in this case) has to be dropped before adding

the appropriate genitive ending. The specific epithets in genitive form are not related to the gender of the genus. Illustrative examples are listed below:

Name of person	Specific epithet	Binomial
Royle	*roylei*	*Impatiens roylei*
Hooker	*hookeri*	*Iris hookeri*
Wallich	*wallichii*	*Euphorbia wallichii*
Sengupta	*senguptae*	*Euphorbia senguptae*
Linnaeus	*linnaei*	*Indigofera linnaei*

(ii) When used in adjectival form the epithet takes its ending from the gender of the genus after adding *-ian* to the name of the person. Illustrative examples are given below:

Name of person—Webb

Genus	Gender	Specific epithet	Binomial
Rosa	Feminine	*webbiana*	*Rosa webbiana*
Rheum	Neuter	*webbianum*	*Rheum webbianum*
Astragalus	Masculine	*webbianus*	*Astragalus webbianus*

2. **Place.** The specific epithet may similarly be formed by using the place name as an adjective, when again the genus determines the ending after the addition of *-ian* or *-ic* and then the relevant gender ending as determined by the genus. Different situations are illustrated below:

Place	Genus	Gender	Specific epithet	Binomial
Kashmir	*Iris*	Feminine	*kashmiriana*	*Iris kashmiriana*
	Delphinium	Neuter	*kashmirianum*	*Delphinium kashmirianum*
	Tragopogon	Masculine	*kashmirianus*	*Tragopogon kashmirianus*
India	*Rosa*	Feminine	*indica*	*Rosa indica*
	Solanum	Neuter	*indicum*	*Solanum indicum*
	Euonymus	Masculine	*indicus*	*Euonymus indicus*

The specific epithet is also formed by adding *-ensis* (for masculine and feminine genera, e.g. *Hedera nepalensis, Rubus canadensis*) or *-ense* (for neuter genera, e.g. *Ligustrum nepalense*) to the place name.

3. **Character.** Specific epithets based on a character of the species are always in adjectival form and derive their gender from the genus. A name based on a white plant part may take the form *alba* (*Rosa alba*), *album* (*Chenopodium album*) or *albus* (*Mallotus albus*). A common epithet used for cultivated plants may similarly take the form *sativa* (*Oryza sativa*), *sativum* (*Allium sativum*) or *sativus* (*Lathyrus sativus*) depending on the gender of the genus to which the epithet is assigned. Some

epithets, however, such as *bicolor* (two-coloured) and *repens* (creeping) remain unchanged, e.g. *Ranunculus repens, Ludwigia repens* and *Trifolium repens*.

4. **Noun in apposition.** A specific epithet may sometimes be a noun in apposition carrying its own gender, and usually in the nominative case. Binomial *Pyrus malus* is based on the Greek name malus for common apple. In *Allium cepa*, similarly cepa is the Latin name for onion.

Both generic name and the specific epithet are underlined when written or typed. When printed, they are in Italics or boldface. After the generic name in a species has been spelled out at least once, if used for other species it may be abbreviated using the initial capital, e.g. *Quercus dilatata, Q. suber, Q. Ilex*, etc. A specific epithet is usually one word but when consisting of two words these must be hyphenated as in *Capsella bursa-pastoris*.

Infraspecific Categories

The names of **subspecies** are **trinomials** and are formed by adding a subspecific epithet to the name of a species., e.g. *Angelica archangelica* ssp. *himalaica*. A variety (**varieta**) within a subspecies may accordingly be **quadrinomial** as in *Bupleurum falcatum* ssp. *eufalcatum* var. *hoffmeisteri*, or it may just be a **trinomial** when no subspecies is recognised within a species as in *Brassica oleracea* var. *capitata*. A **forma** may also be assigned a name in a similar manner, e.g. *Prunus cornuta* forma *villosa*. The formation of the infraspecific epithet follows the same rules as the specific epithet.

V. THE TYPE METHOD

The names of different taxonomic groups are based on the **type method,** by which a certain representative of the group is the source of the name for the group. This representative is called the **nomenclatural type** or simply the **type,** and methodology as **typification**. The type need not be the most typical member of the group, it only fixes the name of a particular taxon and the two are permanently associated. *Mimosa* is the type for family Mimosaceae, but unlike most representatives of the family that have pentamerous flowers, the genus *Mimosa* has tetramerous flowers. The family Urticaceae similarly has *Urtica* as its type. When the originally large family was split into a number of smaller natural families, the name Urticaceae was retained for the group containing the genus *Urtica* since the two cannot be separated. The other splitter groups with family rank got the names Moraceae, Ulmaceae and Cannabinaceae with type genera *Morus, Ulmus* and *Cannabis*, respectively.

The type of a family and the higher groups is ultimately a genus as indicated above. A type of a particular genus is a species, e.g. *Poa pratensis* for *Poa*. The type of name of a species or infraspecific taxon, where it exists, is a single type specimen, preserved in a known herbarium and identified by the place of collection, name of the collector and his collection number. The Code recognises several kinds of types, depending on the way in which a type specimen is selected. These include:

1. **Holotype:** A particular specimen or illustration designated by the author of the species to represent the type of a species. It is now essential to designate a holotype when publishing a new species.

2. **Isotype:** A specimen which is a duplicate of the holotype, collected from the same place, same time and by the same person. Often the collection number is also the same, differentiated as a, b, c, etc.

3. **Syntype:** Any one of the two or more specimens cited by the author when no holotype was designated, or any one of the two or more specimens simultaneously designated as types.

4. **Paratype:** A paratype is a specimen cited in the protologue that is neither the holotype nor an isotype, nor one of the syntypes if two or more specimens were simultaneously designated as types.

5. **Lectotype:** A specimen or other element selected from the original material cited by the author when no holotype was originally selected or when it no longer exists. A lectotype is selected from isotypes, paratypes or syntypes.

6. **Neotype:** A specimen or illustration selected to serve as nomenclatural type as long as all of the material on which the name of the taxon was based is missing. A specimen or an illustration selected when no holotype, isotype, paratype or syntype exists.

7. **Epitype:** A specimen or illustration selected to serve as an interpretative type when the holotype, lectotype or previously designated neotype, or all original material associated with a validly published name, is demonstrably ambiguous and cannot be critically identified for purposes of the precise application of the name of a taxon. When an epitype is designated, the holotype, lectotype or neotype that the epitype supports must be explicitly cited.

In most cases in which no holotype was designated there will also be no paratypes, since all the cited specimens will be syntypes. However, when an author has designated two or more specimens as types, any remaining cited specimens are paratypes and not syntypes.

Topotype is often the name given to a specimen collected from the same locality from which the holotype was originally collected.

When an infraspecific variant is recognised within a species for the first time, it automatically establishes two **infraspecific taxa**. The one which includes the type specimen of the species, must have the same epithet as that of the species, e.g. *Acacia nilotica* ssp. *nilotica*. Such a name is called an

autonym, and the specimen an **autotype**. The variant taxon would have its own holotype and is differentiated by an epithet different from the specific epithet, e.g. *Acacia nilotica* ssp. *indica*.

It must be borne in mind that application of the type method or **typification** is a methodology different from **typology**, which is a concept based on the idea that does not recognise variation within the taxa, and believes that an idealised specimen or pattern can represent a natural taxon. This concept of typology was very much in vogue before Darwin put forward his ideas about variations.

VI. AUTHOR CITATION

For a name to be complete, accurate and readily verifiable it should be accompanied by the name of the author or authors who first published the name validly. The names of the authors are commonly abbreviated, e.g. **Linn.** or **L.** for Carolus Linnaeus, **Benth.** for G. Bentham, **Hook.** for William Hooker, **Hook.f.** for Sir J.D. Hooker (f. stands for filius, the son; J.D. Hooker was son of William Hooker), **R.Br.** for Robert Brown, **Lamk.** for J.P. Lamarck, **DC.** for A.P. de Candolle, **Wall.** for Wallich, **A. DC.** for Alphonse de Candolle, **Scop.** for G.A. Scopoli and **Pers.** for C.H. Persoon.

Single author: The name of a single author follows the name of a species (or any other taxon) when a single author proposed a new name, e.g. *Solanum nigrum* Linn.

Mutiple authors: The names of two or more authors may be associated with a name for a variety of reasons. These different situations are exhibited by citing the name of the authors differently:

1. **Use of *et*:** When two or more authors publish a new species or propose a new name, their names are linked by *et*, e.g. *Delphinium viscosum* Hook.f. *et* Thomson.

2. **Use of parentheses:** The rules of botanical nomenclature specify that whenever the name of a taxon is changed by transfer from one genus to another or by upgrading or downgrading the level of the taxon, the original epithet should be retained. The name of the taxon providing the epithet is termed a **basionym**. The name of the original author or authors whose epithet is being used in the changed name is placed within parentheses, and the author or authors who made the name change outside the parentheses, e.g. *Cynodon dactylon* (Linn.) Pers., based on the basionym *Panicum dactylon* Linn., the original name for the species.

3. **Use of *ex*:** The names of two authors are linked by *ex* when the first author had proposed a name but was validly published only by the second author, the first author failing to satisfy all requirements of the Code, e.g. *Cerasus cornuta* Wall. *ex* Royle.

4. **Use of *in*:** The names of authors are linked using *in* when the first author published a new species or a name in a publication of another author, e.g. *Carex kashmirensis* Clarke *in* Hook.f. Clarke published this new species in the *Flora of British India* whose author was Sir J.D. Hooker.

5. **Use of *emend*:** The names of two authors are linked using *emend*. (*emendavit:* person making the correction) when the second author makes some change in the **diagnosis** or in **circumscription** of a taxon without altering the type, e.g. *Phyllanthus* Linn. *emend*. Mull.

6. **Use of brackets:** Brackets are used to indicate prestarting point author. The generic name *Lupinus* was effectively published by Tournefort in 1719, but as it happens to be earlier than 1753, the starting date for botanical nomenclature based on *Species plantarum* of Linnaeus, the appropriate citation for the genus is *Lupinus* [Tourne.] Linn.

When naming an infraspecific taxon, the authority is cited both for the specific epithet and the infraspecific epithet, e.g. *Acacia nilotica* (Linn.) Del. ssp. *indica* (Benth.) Brenan. In the case of an **autonym**, however, the infraspecific epithet does not bear the author's name since it is based on the same type as the species, e.g. *Acacia nilotica* (Linn.) Del. ssp. *nilotica*.

VII. PUBLICATION OF NAMES

The name of a taxon, when first published, should meet certain requirements so as to become a legitimate name for consideration when deciding on the correct name. A valid publication should satisfy the following requirements:

1. **Formulation:** A name should be properly formulated and its nature indicated by a proper abbreviation after the name of the author:

    (i) *sp. nov.* for *species nova*, a species new to science.

    (ii) *comb. nov.* for *combinatio nova*, a name change involving the epithet of the basionym, name of the original author being kept within parentheses.

    (iii) *nom. nov.* for *nomen novum*, when the original name is replaced and its epithet cannot be used in the new name.

    These abbreviations are, however, used only when first published. In future references these are replaced by the name of the publication, page number and the year of publication for full citation.

2. **Latin diagnosis:** Names of all new species (or other taxa new to science) published 1-1-1935 onwards should have a **Latin diagnosis** (Latin translation of diagnostic features). Full description of the species in any language can accompany the Latin diagnosis. A description in any language, not accompanied by a Latin diagnosis, is allowed for publications before 1-1-1935. For publications before 1-1-1908, an **illustration with analysis** without any accompanying description is valid. Thus description in any language is essential from 1-1-1908 onwards

and this accompanied by a Latin diagnosis from 1-1-1935. For name changes or new names of already known species, a full reference to the original publication should be made.

3. **Typification:** A **holotype** should be designated. Publication on or after 1 January 1958 of the name of a new taxon of the rank of genus or below is valid only when the type of the name is indicated. For the name of a new taxon of the rank of genus or below published on or after 1 January 1990, indication of the type must include one of the words "typus" or "holotypus", or its abbreviation, or its equivalent in a modern language. For the name of a new species or infraspecific taxon published on or after 1 January 1990 whose type is a specimen or unpublished illustration, the herbarium or institution in which the type is conserved must be specified.

4. **Effective publication:** The publication becomes effective by distribution in printed form, through sale, exchange or gift to the general public or at least the botanical institutions with libraries accessible to botanists generally. The publication in newspapers and catalogues (1-1-1953 onwards) and seed exchange lists (1-1-1977 onwards) is not an effective publication. Publication of handwritten material, reproduced by some mechanical or graphic process (**indelible autograph**) before 1-1-1953 is effective.

The date of a name is that of its valid publication. When the various conditions for valid publication are not simultaneously fulfilled, the date is that on which the last is fulfilled. However, the name must always be explicitly accepted in the place of its validation. A name published on or after 1 January 1973 for which the various conditions for valid publication are not simultaneously fulfilled is not validly published unless a full and direct reference is given to the places where these requirements were previously fulfilled.

In order to be accepted a name of a new taxon of fossil plants published on or after 1 January 1996 must be accompanied by a Latin or English description or diagnosis or by a reference to a previously and effectively published Latin or English description or diagnosis.

Subject to ratification by the XVI International Botanical Congress (St Louis, 1999) of a rule already included in the International Code of Botanical Nomenclature (Art. 32.1-2 of the Tokyo Code), new names of plants and fungi will have to be registered in order to be validly published after the 1st of January 2000. After 1 January 2000, when one or more of the other conditions for valid publication have not been met prior to registration, the name must be resubmitted for registration after these conditions have been met. A trial registration has already begun, on a non-mandatory basis, for a two-year period starting 1 January 1998. The co-ordinating centre will be the secretariat of IAPT, currently at the Botanic Garden and Botanical Museum Berlin-Dahlem, Germany. The International Mycological Institute in Egham, UK, has already agreed to act as associate registration centre for the whole of fungi, including fossil fungi.

A correction of the original spelling of a name does not affect its date of valid publication.

VIII. REJECTION OF NAMES

The process of selection of correct name for a taxon involves the identification of **illegitimate names**, those which do not satisfy the rules of botanical nomenclature. Any one or more of the following situations leads to the rejection of a name:

1. *Nomen nudum* (abbreviated *nom. nud.*): A name with no accompanying description. Many names published by Wallich in his *Catalogue*
   (abbreviated *Wall. Cat.*) published in 1812 were *nomen nudum*. These
   were either validated by another author at a later date by providing a
   description (e.g. *Cerasus cornuta* Wall. *ex* Royle) or if by that time the
   name has already been used for another species by some other author,
   the *nomen nudum* even if validated is rejected and a new name has to
   be found (e.g. *Quercus dilatata* Wall., a *nom. nud.* rejected and replaced
   by *Q. himalayana* Bahadur, 1972).
2. Name not effectively published, not properly formulated, lacking typification or without a Latin diagnosis.
3. **Tautonym:** Whereas the Zoological Code allows binomials with identical generic name and specific epithet (e.g. *Bison bison*), such names in
   Botanical nomenclature constitute **tautonyms** (e.g. *Malus malus*) and
   are rejected. The words in the tautonym are exactly identical while
   evidently names such as *Cajanus cajan* or *Sesbania sesban* are not
   tautonyms and thus legitimate. Repetition of a specific epithet in an
   infraspecific epithet does not constitute a tautonym but a legitimate
   autonym (e.g. *Acacia nilotica* ssp. *nilotica*).
4. **Later homonym:** Just as a taxon should have one correct name, the
   Code similarly does not allow the same name to be used for two
   different species (or taxa). Such, if existing, constitute **homonyms**. The
   one published at an earlier date is termed the **earlier homonym** and
   that at a later date as the **later homonym**. The Code rejects later homonyms even if the earlier homonym is illegitimate. *Zizyphus jujuba*
   Lamk., 1789 had long been used as the correct name for the cultivated
   fruit jujube. This, however, was ascertained to be a later homonym of
   a related species *Z. jujuba* Mill., 1768. The binomial *Z. jujuba* Lamk.,
   1789 is thus rejected and jujube correctly named as *Z. mauratiana* Lamk.,
   1789. Similarly, although the earliest name for almonds is *Amygdalus
   communis* Linn., 1753 when transferred to the genus *Prunus* the name
   Prunus communis (Linn.) Archangeli, 1882 for almond became a later
   homonym of *Prunus communis* Huds., 1762 which is a species of plums.

P. communis (Linn.) Archangeli was as such replaced by *P. dulcis* (Mill.) Webb, 1967 as the name for almonds.

5. ***Nomen superfluum*** (abbreviated as ***nom. superfl.***): A name is illegitimate and must be rejected when it was nomenclaturally superfluous when published, i.e., if the taxon to which it was applied, as circumscribed by its author, included the type of a name or epithet which ought to have been adopted under the rules. *Physkium natans* Lour., 1790 thus when transferred to the genus *Vallisneria*, the epithet *natans* should have been retained but de Jussieu used the name *Vallisneria physkium* Juss., 1826 a name which becomes superfluous. The species has accordingly been named correctly as *Vallisneria natans* (Lour.) Hara, 1974. A combination based on a superfluous name is also illegitimate. *Picea excelsa* (Lam.) Link is illegitimate since it is based on a superfluous name *Pinus excelsa* Lam., 1778 for *Pinus abies* Linn., 1753. The legitimate combination under *Picea* is thus *Picea abies* (Linn.) Karst., 1880.

6. ***Nomen ambiguum*** (abbreviated as ***nom. ambig.***): A name is rejected if it is used in a different sense by different authors and has become a source of persistent error. The name *Rosa villosa* Linn. is rejected because it has been applied to several different species and has become a source of error.

7. ***Nomen confusum*** (abbreviated as ***nom. confus.***): A name is rejected if it is based on a type consisting of two or more entirely discordant elements, so that it is difficult to select a satisfactory lectotype. The characters of the genus *Actinotinus*, for example, were derived from two genera *Viburnum* and *Aesculus*, owing to the insertion of the inflorescence of *Viburnum* in the terminal bud of an *Aesculus* by a collector. The name *Actinotinus* must, therefore, be abandoned.

8. ***Nomen dubium*** (abbreviated as ***nom. dub.***): A name is rejected if it is dubious, i.e., it is of uncertain application because it is impossible to establish the taxon to which it should be referred. Linnaeus (1753) attributed the name *Rhinanthus crista-galli* to a group of several varieties, which he later described under separate names, rejecting the name *R. crista-galli* Linn. Several later authors, however, continued to use this name for diverse occasions until Schwarz (1939) finally listed this as ***Nomen dubium***, and the name was finally rejected.

9. **Name based on monstrosity:** A name must be rejected if it is based on a monstrosity. The generic name *Uropedium* Lindl., 1846 was based on a monstrosity of the species now referred to as *Phragmidium caudatum* (Lindl.) Royle, 1896. The generic name *Uropedium* Lindl. must, therefore, be rejected. The name *Ornithogallum fragiferum* Vill., 1787 is likewise based on a monstrosity and thus should be rejected.

A legitimate name or epithet, however, must not be rejected merely because it is inappropriate or disagreeable, or because another is preferable or better known, or because it has lost its original meaning. The name *Scilla peruviana* must not be rejected simply because the species does not grow in Peru but is found in the Mediterranean region.

IX. PRINCIPLE OF PRIORITY

The principle of priority is concerned with the selection of a single correct name for a taxonomic group. After identifying **legitimate** and **illegitimate** names, and rejecting the latter, a correct name has to be selected from among the legitimate ones. If more than one legitimate names are available for a taxon, the correct name is the earliest legitimate name in the same rank. For taxa at the species level and below the correct name is either the earliest legitimate name or a combination based on earliest legitimate **basionym**, unless the combination becomes a tautonym or later homonym, rendering it illegitimate. The following examples illustrate the **principle of priority**:

1. The three commonly known binomials for the same species of *Nymphaea* are *N. nouchali* Burm.f., 1768, *N. pubescence* Willd., 1799 and *N. torus* Hook.f. *et* T., 1872. Using the priority criterion, *N. nouchali* Burm.f. is selected as the correct name as it bears the earliest date of publication. The other two species are regarded as **synonyms**. The citation is written as:

   Nymphaea nouchali Burm.f., 1768
   N. pubescence Willd., 1799
   N. torus Hook.f. *et* T., 1872

2. Loureiro described a species under the name *Physkium natans* in 1790. It was subsequently transferred to the genus *Vallisneria* by A. L. de Jussieu in 1826, but unfortunately he ignored the epithet *natans* and used a binomial *Vallisneria physkium*, a **superfluous name**. Two Asiatic species with independent typification were described subsequently under the names *V. gigantea* Graebner, 1912 and *V. asiatica* Miki, 1934. Hara on making a detailed study of Asiatic specimens concluded that all these names are synonymous, and also that *V. spiralis* Linn. with which most of the Asiatic specimens were identified does not grow in Asia. As no legitimate combination based on *Physkium natans* Lour. existed, he made one—*V. natans* (Lour.) Hara—in 1974. Thus the correct name of the species in this case is the most recent name, but it is based on the earliest **basionym**. The synonymy would be cited as under:

   Vallisneria natans (Lour.) Hara, 1974
   Physkium natans Lour., 1790—Basionym
   V. physkium Juss., 1826—*nom. superfl.*
   V. gigantea Graebner, 1912

V. asiatica Miki, 1934

V. spiralis auct. (non Linn., 1753)

It must be noted that *Physkium natans* and *Vallisneria physkium* are based on the same type as the correct name *V. natans* and are thus known as **nomenclatural synonyms** or **homotypic synonyms**. These three would remain together in all citations. The other two names *V. gigantea* and *V. asiatica* are based on separate types and may or may not be regarded as synonyms of *V. natans* depending on taxonomic judgement. Such a synonym, which is based on a type different from the correct name, is known as a **taxonomic synonym** or **heterotypic synonym**.

3. The common apple was first described by Linnaeus under the name *Pyrus malus* in 1753. The species was subsequently transferred to the genus *Malus* but the combination *Malus malus* (Linn.) Britt., 1888 cannot be taken as the correct name since it becomes a tautonym. Two other binomials under *Malus* available for apple include *M. pumila* Mill., 1768 and *M. domestica* Borkh., 1803; the former being the earlier of the two is selected as the correct name and the citation written as:

Malus pumila Mill., 1768

Pyrus malus Linn., 1753

M. domestica Borkh., 1803

M. malus (Linn.) Britt., 1888—Tautonym

Although the earliest name *Pyrus malus* is perfectly legitimate, since the species is now placed in the genus *Malus* it cannot serve as a basionym for the correct name since *Malus malus* is a tautonym.

4. Almond was first described by Linnaeus under the name *Amygdalus communis* in 1753. Miller described another species under the name *A. dulcis* in 1768. The two are now regarded as synonymous. The genus *Amygdalus* was subsequently merged with the genus *Prunus* and the combination *Prunus communis* (Linn.) Archangeli made in 1882 based on the earlier name *Amygdalus communis* Linn. It was discovered by Webb that the binomial *Prunus communis* had already been used by Hudson in 1762 for a different species rendering *P. communis* (Linn.) Archangeli a **later homonym** which had consequently to be rejected. Webb accordingly used the next available **basionym** *Amygdalus dulcis* Mill., 1768 and made a combination *Prunus dulcis* (Mill.) Webb, 1967 as the correct name for almond. Another binomial, *Prunus amygdalus* Batsch, 1801, cannot be taken up as it ignores the earlier epithets. The citation for almond would thus be:

Prunus dulcis (Mill.) Webb, 1967

Amygdalus dulcis Mill., 1768—basionym

A. communis Linn., 1753

Prunus communis (Linn.) Arch., 1882 (non Huds., 1762)

P. amygdalus Batsch, 1801

Limitations to the principle of priority

Application of the principle of priority has the following limitations:

1. **Starting dates:** The principle of priority starts with the *Species plantarum* of Linnaeus published 1-5-1753. The starting dates for different groups include:

    Seed plants, Pteridophytes........ 1-5-1753
    Mosses...1-1-1801
    Fungi... 31-12-1801
    Algae, Hepaticae, Sphagnaceae,
    lichens and fossils.......................31-12-1820

    The publications before these dates for respective groups are ignored for deciding the priority.

2. **Only up to family rank:** The principle of priority is not applicable above the family rank.

3. **Not outside the rank:** In choosing a correct name for a taxon, names or epithets available at that rank are to be considered. Only when a correct name at that rank is not available, can a combination be made using the epithet from another rank. Thus at the level of section the correct name is *Campanula* sect. *Campnopsis*, 1810 but when upgraded as a genus the correct name is *Wahlenbergia* Schrad, 1821. The following names are synonyms:

    > *Lespedza eriocarpa* DC. var. *falconeri* Prain, 1897
    > *L. meeboldii* Schindler, 1911
    > *Campylotropis eriocarpa* var. *falconeri* (Prain) Nair, 1977
    > *C meeboldii* (Schindler) Schindler, 1912

    The correct name at the species level under the genus *Campylotropis* would be *C. meeboldii* ignoring the earlier epithet at the varietal level. If treated as a variety the correct name would be *C. eriocarpa* var. *falconeri*, based on the earliest epithet at that rank. Under the genus *Lespedza*, at the species level the correct name would be *L. meeboldii* whereas at the varietal level, it would be *L. eriocarpa* var. *falconeri*.

4. *Nomina conservanda* (abbreviated as **nom. cons.**): Strict application of the principle of priority has resulted in numerous name changes. To avoid name changes of well-known families or genera especially those containing many species, a list of conserved generic and family names has been prepared and published in the Code with relevant changes. Such *nomina conservanda* are to be used as correct names replacing the earlier legitimate names, which are rejected and constitute *nomina rejicienda* (abbreviated **nom. rejic.**). The family name Theaceae D. Don, 1825 is thus conserved against Ternstroemiaceae Mirbe, 1813. The genus *Sesbania* Scop., 1777 is conserved against *Sesban* Adans., 1763 and *Agati* Adans., 1763.

Conservation of Names of Species

In spite of several protests from agricultural botanists and horticulturists, who were disgusted with frequent name changes due to strict application of the principle of priority, taxonomists for a long period did not agree upon conserving names at the species level. The mounting pressure and the discovery that *Triticum aestivum* was not the correct name of common wheat, prompted taxonomists to agree at the **Sydney Congress** in 1981 upon the provision to conserve names of species of major economic importance. As a result, *Triticum aestivum* Linn. was the first species name conserved at the **Berlin Congress** in 1987 and was published in the subsequent Code in 1988. Another species name conserved at the same time was *Lycopersicon esculentum* Mill.

Linnaeus described two species, *Triticum aestivum* and *T. hybernum* in his *Species plantarum*, both bearing the same date of publication, 1753. According to the rules of nomenclature when two species with the same date of publication are united, the author who unites them first has the choice of selecting the correct binomial. For a long time it was known that the first persons to unite the two species were Fiori and Paoletti in 1896 who selected *T. aestivum* Linn. as the correct name. It was pointed out by Kergnelen (1980), however, that the first person to unite these two species was actually Merat (1821) and he had selected *T. hybernum* Linn. and not *T. aestivum*. This discovery led to the danger of *T. aestivum* Linn. being dropped in favour of *T. hybernum* Linn. A proposal for conserving the name *T. aestivum* Linn. was thus made by Hanelt and Schultze-Motel (1983), and being the number one economic plant this was accepted at the Berlin Congress, removing any further danger to the name *Triticum aestivum* Linn.

P. Miller in 1768 proposed a new name, *Lycopersicon esculentum*, for tomato, a species described earlier by Linnaeus (1753) as *Solanum lycopersicum*. Karsten (1882) made the name change *Lycopersicum lycopersicum* (Linn.) Karst., retaining the epithet used by Linnaeus, but since the name became a **tautonym** it was not considered the correct name for tomato. Nicolson (1974) suggested an orthographic correction *Lycopersicon lycopersicum* (Linn.) Karst., suggesting that *Lycopersicon* and *lycopersicum* are orthographic variants. Since the name *Lycopersicon lycopersicum* was no longer a tautonym, it was accepted as the correct name. But since *Lycopersicon esculentum* Mill., 1768 was a more widely known name, a proposal for its conservation was made by Terrel (1983) and accepted at the Berlin Congress along with that of *Triticum aestivum* Linn. A list is given below of species names which have been declared nomina conservanda (each name followed by the (=) sign indicating taxonomic synonym or a (= =) sign indicating nomenclatural synonym and then the binomial against which it has been conserved) in the 1993 Code (published in 1994):

Bactris gasipaes Kunth, 1816.

(=)

Martinezia ciliata Ruiz et Pav., 1798 (*Bactris ciliata* (Ruiz et Pav.) Mart.).

Erica carnea L., 1753

(=)

Erica herbacea L., 1753.

Fraxinus angustifolia Vahl, 1804

(=)

Fraxinus rotundifolia Mill., 1768.

Lycopersicon esculentum Mill., 1768

(= =)

Lycopersicon lycopersicum (L.) H. Karsten, 1882

Triticum aestivum L., 1753

(=)

Triticum hybernum L., 1753.

X. NAMES OF HYBRIDS

Hybridity is indicated by the use of the multiplication sign, 'x' or by the addition of the prefix 'notho-' to the term denoting the rank of the taxon. A hybrid between named taxa may be indicated by placing the multiplication sign between the names of the taxa; the whole expression is then called a **hybrid formula**:

1. *Agrostis* × *Polypogon*.
2. *Agrostis stolonifera* × *Polypogon monspeliensis*
3. *Salix aurita* × *S. caprea*

It is usually preferable to place the names or epithets in a formula in alphabetical order. The direction of a cross may be indicated by including the sexual symbols ({♀}: female; {♂}: male) in the formula, or by placing the female parent first. If a non-alphabetical sequence is used, its basis should be clearly indicated.

A hybrid may either be **interspecific** (between two species belonging to the same genus) or **intergeneric** (between two species belonging to two different genera). A binary name may be given to the interspecific hybrid or nothospecies (if it is self-perpetuating and/or reproductively isolated) by placing a cross sign (if the mathematical sign is available it should be placed immediately before the specific epithet, otherwise 'x' in lower case may be used with a gap) before the specific epithet as in the following cases (hybrid formula may be added within the parentheses if the parents are established):

1. *Salix* ×*capreola* (*S. aurita* × *S. caprea*)
   or *Salix* x *capreola* (*S. aurita* × *S. caprea*)
2. *Rosa* ×*odorata* (*R. chinensis* × *R. gigantea*)
   or *Rosa* x *odorata* (*R. chinensis* × *R. gigantea*)

The variants of interspecific hybrids are named **nothosubspecies** and **nothovarieties**, e.g. *Salix rubens* nothvar. *basfordiana*.

For an intergeneric hybrid, if given a distinct generic name, the name is formed as a **condensed formula** by using the first part (or whole) of one parental genus and last part (or whole) of another genus (but not the whole of both genera). A cross sign is placed before the generic name of the hybrid, e.g. × *Triticosecale* from *Triticum* and *Secale*, × *Pyronia* from *Pyrus* and *Cydonia*. The names may be written as under:

1. × *Triticosecale* (*Triticum* × *Secale*)
2. × *Pyronia* (*Pyrus* × *Cydonia*)

The nothogeneric name of an intergeneric hybrid derived from four or more genera is formed from the name of a person to which is added the termination **-ara**; no such name may exceed eight syllables. Such a name is regarded as a condensed formula: × *Potinara* (*Brassavola* × *Cattleya* × *Laelia* × *Sophronitis*).

The nothogeneric name of a trigeneric hybrid is either (a) a condensed formula in which the three names adopted for the parental genera are combined into a single word not exceeding eight syllables, using the whole or first part of one, followed by the whole or any part of another, followed by the whole or last part of the third (but not the whole of all three) and, optionally, one or two connecting vowels, or (b) a name formed like that of a nothogenus derived from four or more genera, i.e., from a personal name to which is added the termination **-ara:** × *Sophrolaeliocattleya* (*Sophronitis* × *Laelia* × *Cattleya*); When a nothogeneric name is formed from the name of a person by adding the termination **-ara**, that person should preferably be a collector, grower, or student of the group.

A binomial for the intergeneric hybrid may similarly be written as under:

× *Agropogon lutosus* (*Agrostis stolonifera* × *Polypogon monspeliensis*)

It is important to note that a binomial for an interspecific hybrid has a cross before the specific epithet, whereas in an intergeneric hybrid it is before the generic name. Since the names of nothogenera and nothotaxa with the rank of a subdivision of a genus are condensed formulae or treated as such, they do not have types.

Since the name of a nothotaxon at the rank of species or below has a type, statements of parentage play a secondary part in determining the application of the name.

The grafts between two species are indicated by a plus sign between two grafted species as, for example, *Rosa webbiana* + *R. floribunda*.

XI. NAMES OF CULTIVATED PLANTS

The names of cultivated plants are governed by the International Code of Nomenclature for Cultivated Plants (**ICNCP**), last published in 1995 (Trehane et al.). Most of the rules are taken from ICBN with additional recognition of a rank **cultivar** (abbreviated **cv.**) for cultivated varieties. The name of a cultivar is not written in Italics, starts with a capital letter, and is not a Latin but rather a common name. It is either preceded by **cv.** as in *Rosa floribunda* cv. Blessings or simply within single quotation marks, e.g. *Rosa floribunda* 'Blessings'. Cultivars may also be named directly under a genus (e.g. *Hosta* 'Decorata'), under a hybrid (e.g. *Rosa* × *paulii* 'Rosea') or directly under a common name (e.g. Hybrid Tea Rose 'Red Lion'). The correct nothogeneric name for plants derived from the *Triticum* × *Secale* crosses is × *Triticosecale* Wittmack ex A. Camus. No correct name at the species level is available for the commonest crop triticales. It is recommended that crop triticales be named by appending the cultivar name to the nothogeneric name, e.g., × *Triticosecale* 'Newton'. Since 1-1-1959 new cultivar names should have a description published in any language and these names must not be the same as the botanical or common name of a genus or a species. Thus cultivar names 'Rose', 'Onion' etc. are not permitted as the name of a cultivar. It is recommended that cultivar names be registered with **registering authorities** to prevent duplication or misuse of cultivar names. Registering authorities exist separately for Roses, Orchids and several other groups or genera.

XII. DRAFT BIOCODE (Attempt Towards a Unified Code)

Biology as a science is unusual in that the objects of its study (as indicated earlier) can be named according to five different Codes of nomenclature: International Code of Zoological Nomenclature(**ICZN**) for animals (Ride et al., 1985), International Code of Botanical Nomenclature (**ICBN**) for plants (Greuter et al., 1994), International Code for the Nomenclature of Bacteria (ICNB) now called Bacteriological Code (**BC**) for bacteria (Lapage et al., 1992), International Code of Nomenclature for Cultivated Plants (**ICNCP**, Trehane et al., 1995), and International Code of Virus Classification and Nomenclature (draft, currently being developed from the most recent Rules of Virus Classification and Nomenclature by the International Committee for the Taxonomy of Viruses (**ICTV**)). For the general user of scientific names of organisms, there is thus inherent confusion in many aspects of this situation: different sets of rules have different conventions for citing names, provide for different forms for names at the same rank, and, although primarily each is based on priority of publication, differ somewhat in how they determine the choice of correct name. This diversity of Codes can also create more serious problems as, for example, in the determination of which Code to follow for those organisms that are not clearly plants, animals or bacteria, the so-called ambiregnal organisms, or those whose current genetic affinity may be well established but whose traditional treatment has been in a different group (e.g. the cyanobacteria). Moreover, the development of electronic information retrieval, by often using scientific names without clear taxonomic context, accentuates the problem of divergent methods of citation and makes homonymy between, for example, plants and animals a source of trouble and frequently confusion.

The desirability of seeking some harmonisation of all biological codes has been appreciated for some time (see Hawksworth, 1995) and an exploratory meeting on the subject was held at Egham, UK in March 1994. Recognising the crucial importance of scientific names of organisms in global communication, these decisions included not only agreement to take steps to harmonise terminology and procedures, but also the desirability of working towards a unified system of biological nomenclature. The **Draft BioCode** is the first public expression of these objectives. The first draft was prepared in 1995. After successive reviews the fourth draft, named **Draft BioCode (1997)**, prepared by the International Committee for Bionomenclature and published by Greuter et al. is now available on the web: (http://www. rom.on.ca/biodiversity/biocode/biocode.html) from the Royal Ontario Museum.

Preamble

1. Biology requires a precise, coherent and simple system for the naming of organisms used internationally, dealing both with the terms which denote the ranks of taxonomic groups and with the scientific names which are applied to the individual taxonomic groups of organisms.
2. The provisions of this Code shall apply to names of all kinds of **non-viral organisms**, whether fossil or non-fossil, and to some fossil traces of organisms, published and established on or after 1 January 2000, and shall govern the choice of name when these names compete among themselves or with earlier names. They shall also, and without limitation of date, provide, in the interest of nomenclatural stability and security, for the protection, conservation, or suppression of all such names, as well as for their correct form and spelling.
3. Names of non-viral organisms that have been established (i.e., validly published or become available) prior to 1 January 2000 and are not yet covered by adopted Lists of Protected Names are in all other respects (including their subsequent typification) governed by the various existing Codes depending on the accepted taxonomic position of their type.
4. Special provisions apply to the nomenclature of particular groups of organisms, notably viruses and cultivated plants.
5. Separate rules for virus nomenclature, contained in The International Code of Virus Classification and Nomenclature, have been established in conformity with Principles I & V of this Code and with the thrust of many of its rules. Because names of virus species do not have the binominal form required under this Code, and names of virus taxa in

other recognised ranks have mandatory terminations according to rank, provisions of this Code proscribing these terminations for non-virus taxa ensure that the names of viruses and other organisms cannot conflict.

6. The nomenclature of cultivated plants follows the provisions of this Code, insofar as these provisions are applicable, but the naming of distinguishable groups of plants whose origin or selection is primarily due to the intentional actions of mankind follow the supplementary provisions of the International Code of Nomenclature for Cultivated Plants.

In this Code, the term 'fossil' is applied to a taxon when its name is based on a fossil type and the term 'non-fossil' is applied to a taxon when its name is based on a non-fossil type.

Principles

BioCode has three additional principles (Nos. II, VII, VIII) compared to the Botanical Code with minor changes in the other six, more so in Principle I.

Principle I: The BioCode governs the formation and choice of scientific names of known taxa but not the definition of the taxa themselves. Nothing in this Code may be construed to restrict the freedom of taxonomic thought or action.

Principle II: Scientific nomenclature of organisms builds upon the Linnaean system of binary names for species.

Principle III: The application of names of taxa is determined by means of name-bearing types, although application of this principle is not universal at supra-familial ranks.

Principle IV: Nomenclature of a taxon is based upon precedence of publication, although application of this principle is not mandatory at all ranks.

Principle V: Each taxon in the family-group, genus-group or species-group with a particular circumscription, position and rank has only one accepted name, except as may be specified in earlier Codes.

Principle VI: Scientific names of taxa are by convention treated as if they were Latin, regardless of their derivation.

Principle VII: The only proper reasons for changing a name are either a change in the circumscription, position or rank of the taxon, resulting from adequate taxonomic study, or the promotion of nomenclatural stability.

Principle VIII: In the absence of a relevant rule or where the consequences of rules are doubtful, established custom is followed.

Principle IX: The rules of nomenclature are retroactive unless expressly limited (but see Pre. 2-3).

Salient Features of Draft BioCode

Largely on the pattern of the the Botanical Code the salient features of this Draft BioCode include:

(i) **General points:** No examples are listed. Notes and Recommendations have also been omitted at the present stage, although some will no doubt be needed. A very considerable number of articles and paragraphs have been dropped; the Draft BioCode has only **41 Articles**, whereas the Tokyo Code has **74.**

(ii) **Ranks, priority:** The present ranks of the Botanical Code are maintained in the Draft BioCode, and a few tentatively added: **domain** (above kingdom), in use for the pro-/eukaryotes, **superfamily** (in widespread use in zoology), and the option of adding the prefix **super-** to rank designations that are not already prefixed (Art. 3-4). The principal ranks of taxa in descending sequence are: **kingdom, phylum, class, order, family, genus**, and **species.** Secondary ranks of taxa, when required, include in descending sequence: **domain** above kingdom, **superfamily** above family, subfamily and tribe between family and genus, subgenus, section and series between genus and species, and subspecies, variety and form below species. When an even greater number of ranks of taxa is desired, the terms for these are made by adding the prefixes **super-** or **sub-**, or **infra-** (infra- being below sub-) to non-prefixed terms denoting principal or secondary ranks. Superspecies or infraspecies are permitted but not infrasubspecies. Throughout this Code the term '**suprafamilial**' refers only to ranks above the family group; the phrase '**family group**' refers to the ranks of superfamily, family and subfamily; '**subdivision of a family**' only to taxa of a rank between family group and genus group; '**genus group**' refers to the ranks of genus and subgenus; '**subdivision of a genus**' only to taxa of a rank between genus group and species group; '**species group**' to the ranks of species and subspecies; and the term '**infrasubspecific**' refers to ranks below the species group.The principle of mandatory precedence (priority) is to operate only within three of them—**family group, genus group**, and **species group**—because vertical transfers of names across the boundaries of the groups is to be precluded. Both features would be major innovations for botanical nomenclature.

(iii) **Co-ordinate status:** It is proposed that the rule presently prevailing in zoology be extended to botany and bacteriology. This would mean that in the family group, genus group and species group, establishment of any name will automatically establish co-ordinate names, the same authorship and date, at all other ranks of the same group. This rule, which would not, of course operate retroactively, would replace the present autonym rule in botany, and differ from it in two major respects: (a) the date of establishment of the 'autonym equivalent' would usually be earlier (and more easily ascertained) and (b) the 'autonym equivalent' would be established in an upward as well as downward direction in the taxonomic hierarchy, e.g. the establishment of the name of a new subspecies would, at the same time, establish the same name at species rank. Introducing co-ordinate status in the genus group has one major consequence: since any new subgeneric name will simultaneously be established at generic rank, its epithet must have the same form as a generic name and no longer can be a plural adjective, as is currently permitted under the Botanical Code. This rule, concerning the form of names, should logically be retroactive and, if so, would lead to the disestablishment (devalidation) of former subgeneric names of which the epithets (contrary to Rec. 21B.1 of the Tokyo Code) are adjectival. Negative effects of this rule, if any, might be minimised by a minor change, whereby such rather than losing their nomenclatural status, would remain valid but become unranked (and infra-subgeneric).

(iv) **Publication:** Some possible innovations, to account for recent progress in publication technology, have been tentatively incorporated into Art.5.2. They would not of course be retroactive. Publication, under this Code, is defined as distribution of text or images (but not sound) in multiple identical, durable and unalterable copies, in a way that makes it generally accessible as a permanent public record to the scientific community, be it through sale or exchange or gift, and subject to the restrictions and qualifications in the present Article. The distribution of text in encrypted (including digitised) form, in particular the dissemination of text or images solely through erasable electronic support or through electronic communication networks, is not publication. Similarly, communication of the text or images at a public meeting, reproduction of handwritten material in facsimile, e.g. by print, photostat or microfilm, or the distribution of films or photographs of text or images do not constitute publication.

(v) **Establishment of names:** Establishment (valid publication) under the BioCode includes registration as a last step after fulfilment of the present requirements for valid publication. This is nothing new for botanists, being already foreshadowed in the Tokyo Code (Art.32.1-2, 45.2), and an analogous provision requiring indexing by the Zoological Record within five years of publication was included in the draft of the 4th edition of the ICZN. Procedures and mechanisms of registration are yet to be worked out and will be detailed in a special Annexure; these may well be to some extent independent for the various major groups of organisms. Ultimate responsibility for the registration system is assigned to the International Committee on Bionomenclature, but international disciplinary organisations such as the IAPT, although not now explicitly mentioned, are likely to play an active role in the registration of names of taxa. At present, the requirement of Latin descriptive matter for the validation of names of new taxa (if non-fossil) is a unique feature of the Botanical Code. The Draft BioCode (Art. 8.2) opts for a compromise between zoology (any language) and botany (Latin only), and follows the solution pioneered by palaeobotany in that a Latin or an English description is currently required for publication of names of plant fossils. The name of a taxon consisting of the name of a genus combined with one epithet is termed a **binomen**, the name of a species combined with an infraspecific epithet is termed a **trinomen**; binomina or trinomina are also termed combinations.

(vi) **Limitations of precedence (priority):** Adopted lists of names in current use, a much debated issue in botanical nomenclature, would become a newly available option, analogous to what the draft version of the next Zoological Code proposes to rule. For the conservation of names, rank limitations would be abolished, by analogy to the current Zoological Code and as a logical consequence of co-ordinate status of future names within rank groups. The difference with respect to the present situation in botany is in fact minimal, since limitation of precedence makes sense only in rank groups with mandatory precedence. Conservation and rejection procedures would remain largely the same as at present. The botanical process of sanctioning concerns old names only and need not be provided for in a future BioCode.

(vii) **Homonymy:** The major change with respect to the homonymy rule would be that, in future, it would operate across the kingdoms. In order that this provision be applicable, it is necessary that lists of established generic names of all organisms be publicly available, ideally in electronic format; most such, apparently, already exist, but are not yet generally accessible. A list of across-kingdom generic homonyms in current use is being prepared, and, as a next step, a list of binomina in the corresponding genera is planned, so that future workers may avoid the creation of new (illegal) homonymous binomina. Existing across-kingdom homonyms would not lose their status of acceptable names, but would be flagged for the benefit of biological indexers and users of indexes. Existing names are not affected by the proposed rules.

(viii) **Secondary Homonymy:** 'Secondary Homonymy' is the term given by the ICZN to situations in which species-group names established for different nominal taxa (i.e., taxa based on different types) under different generic names are brought together under the same generic name. The zoological practice is to give precedence to the first published name regardless of the date upon which the names are brought into homonymy by taxonomic decision. Botanical practice does not distinguish 'secondary homonymy' and considers that a homonym would only be created with the publication of the 'new combination', the binomen in the genus into which the species are being brought together. The BioCode follows botanical practice in this regard, restricting precedence to the binomina per se, so that an established name can never be altered as a result of a later taxonomic decision.

(ix) **Spelling and gender of names:** Lively discussions are taking place among zoologists, aiming at the abolishment of gender of generic names and the maintenance of the original or a later termination of adjectival epithets upon transfer. Essentially this would remove the long-standing provision in all three Codes, retained in the Draft BioCode (Prin. VI), that scientific names are Latin or deemed to be Latin. This might ultimately result (taking a zoological example) in *Passer domesticus* (L.), based on *Fringilla domestica* L., having to become known as *Passer domestica*. An alternative might be the provision of foolproof recipes to users and inventors of names at three levels: (1) authoritative guidance on the appropriate gender of generic names—already present for a substantial share of botanical names in NCU-3 (Greuter et al., 1994); (2) similar guidance on the appropriate form, spelling and declination of epithets and word elements used in their formation; and (3), perhaps somewhat less urgently, guidelines on the appropriate genitive singular termination to be used in compounding information of suprageneric names. For future names, the registration procedure would offer an excellent opportunity to prevent incorrect usage of gender, or non-standard spellings, from spreading.

(x) **Author citation:** The Draft BioCode signals a departure from the botanical tradition of laying great emphasis on the use of author citations, even in contexts where such citations are neither informative nor really appropriate. This may be a timely change, since attitude is showing signs of cracking (Garnock-Jones and Willis, 1996). Art. 40.1 is so worded as to reflect this new attitude.

(xi) **Ambiregnal organisms:** While many of the provisions of the BioCode will come as a relief to workers in ambiregnal groups, they will not completely solve their problems. Inevitably some rules will remain that are different for different groups of organisms, however defined. Borderline problems are notoriously difficult to solve, and are in fact insoluble unless and until a consensus is reached, among workers in the groups concerned, as to which is the appropriate borderline. As experience tells, such difficulties are surmountable if they can be dealt with under a single Code: there has never been a problem, under the Botanical Code, in delimiting fungi from algae, algae from other plants, or fossil from non-fossil taxa, and there used to be no problems with the 'blue-green algae' so long as bacteria and algae were dealt with under the same Code. It will be the task and privilege of a future BioCode to define which rules apply to dinophytes and dinoflagellates, to euglenids, trichomonads and trypanosomes.

(xii) **Hybrids:** The Appendix for Hybrids in the Botanical Code is replaced by a single Article in the Draft BioCode. This extreme simplification should in no way disrupt present and future usage of hybrid designations, but has some philosophical changes as its basis. Most importantly, taxonomy and nomenclature are disentangled, in conformity with Principle I: nothing remains of the former statement on appropriate rank, or of the requirement of a single hybrid taxon per hybrid combination. The condensed formulae designating intergeneric hybrids are restricted to usage as surrogates of generic names in the formation of binomina. The danger that, in view of that quasi-generic function, they might have to be considered for purposes of homonymy (and thus indexed) has been avoided by the proposed convention of considering the multiplication sign as part of these 'names'.

(xiii) **Special topics:** Art.25.1 endeavours to introduce a clearer definition of what the Botanical Code calls 'descriptive names', at the suprafamilial ranks. Such names being generally used in zoology much more widely than in botany, the clarification was needed. As worded, it appropriately reflects current botanical and zoological practice.

It was originally resolved to make BioCode operative from 1 January 2000, but the present draft code has left it open to 1 January, 200n, depending upon the progress of the project. This was mainly done in light of concerns that implementation would be effected with undue haste.

4

Process of Identification

Recognising an unknown plant is an important constituent taxonomic activity. A plant specimen is identified by comparison with already known herbarium specimens in a herbarium, and by utilising available literature. Since the bulk of our plant wealth grows in areas far removed from the centres of botanical research and training, it becomes imperative to collect a large number of specimens on each outing. For proper description and documentation of these specimens, these have to be suitably prepared for incorporation and permanent storage in a herbarium. This goes a long way in compiling floristic accounts of the different regions of the world.

I. SPECIMEN PREPARATION

A specimen meant for incorporation in a herbarium needs to be carefully collected, pressed, dried, mounted and finally properly labelled, so that it can meet the demands of rigorous taxonomic activity. Specimens, properly prepared, can retain their essential features for a very long period. Such specimens can prove immensely useful for future scientific studies including compilation of floras, taxonomic monographs and, in some cases, even experimental studies, since the seeds of several species can remain viable for many years even in dry herbarium specimens.

Fieldwork

The fieldwork of specimen preparation involves plant collection, pressing and partial drying of the specimens. The plants are collected for various purposes: building new herbaria or enriching older ones, compilation of floras, material for museums and classwork, ethnobotanical studies, and introduction of plants in gardens. In addition, bulk collections are done for trade and drug manufacture. Depending on the purpose, resources, proximity of the area and duration of studies, fieldwork may be undertaken in different ways:

1. **Collection trip:** Of short duration, usually one or two days, to a nearby place, for brief training in fieldwork, vegetation study and plant collection by groups of students.
2. **Exploration:** Repeated visits to an area in different seasons, for a period of a few years, for intensive collection and study, aimed at compilation of floristic accounts.
3. **Expedition:** Undertaken to remote and difficult areas, to study the flora and fauna, and usually takes several months. Most of our early information on Himalayan flora and fauna has been the result of European and Japanese expeditions.

Equipment: The equipment for fieldwork may involve a long list, but the items essential for collection include plant press, field notebook, bags, vasculum, pencil, cutter, pruning shears, knife and a digging tool such as trowel or pickaxe (Fig. 4.1).

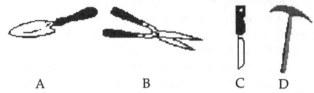

A   B   C   D

Fig. 4.1 Common implements helpful in collection. A—Trowel. B—Pruning shears. C—Knife. D—Pickaxe.

1. **Plant press:** A plant press consists of two wooden, plywood or wire mesh planks, each 12 inches × 18 inches (30 cm × 45 cm), between which are placed corrugated sheets, blotters and newspaper sheets (Fig. 4.2). Two straps, chains or belts are used to tighten the press. **Corrugated sheets** or ventilators are made of cardboard, and help ventilation and the consequent drying of specimens. The ducts of the corrugated sheet run across and not lengthwise to afford shorter distance and larger number of ducts.

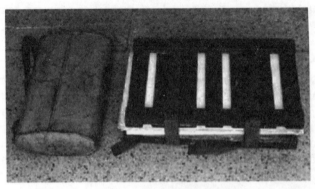

Fig. 4.2 A plant press containing pressed specimens and a vasculum alongside.

The plant press carried in the field and called a **field press** is light weight and generally has one corrugated sheet alternating with one folded blotter containing ten folded newspaper sheets, one meant for each specimen.

The plant press used for subsequent pressing and drying of specimens, kept at the base camp or the organisation, is called the **drying press**. It is much heavier and has an increased number of corrugated sheets, one alternating each folded blotter containing one folded newspaper. In countries such as India which use thick coarse paper for newsprint, blotters can be dispensed with, in at least subsequent changes, as the paper soaks sufficient moisture and serves the purpose of blotters as well.

2. **Field notebook:** Field notebook or field diary is an important item for a collector. A well-designed field notebook (Fig. 4.3) has numbered sheets with printed Proforma for entering field notes such as scientific name, family, vernacular name, locality, altitude, date of collection and for recording any additional data collected in the field. The multiple detachable slips at the lower end of the sheet, separated by perforated lines and bearing the serial number of the sheet, can be used as tags for multiple specimens of a species collected from a site, and serve as ready reference to the information recorded in the field notebook. The number also serves as the collection number for the collector.

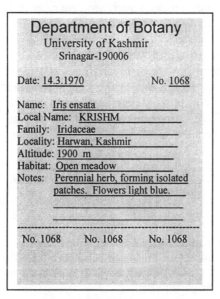

Fig. 4.3   A sheet from the field notebook with relevant entries.

3. **Vasculum:** This is a metallic box with a tightly fitted lid and with a shoulder sling. It is used to store specimens temporarily before pressing, and also to store bulky parts and fruits. It is generally painted white to

deflect heat and affords easy detection when left in the field. Being bulky, the vasculum is commonly substituted by a **polythene bag,** which is almost weightless. A number of polythene bags can be carried for easy storage, as these can be readily made airtight using a rubber band and, as such, the plants retain freshness for many hours.

Collection: The specimen collected should be as complete as possible. Herbs, very small shrubs, as far as possible should be collected complete, in flowering condition, along with leaves and roots. Trees and shrubs should be collected with both vegetative and flowering shoots, to enable the representation of both leaves and flowers. All information concerning the plant should be recorded in the field notebook and a tag from the sheet attached to the concerned specimen. It is advisable to collect a few specimens of each species from the site, to ensure that reserve specimens are available if one or more get destroyed, and also to ensure that duplicates can be deposited in different herbaria, when finally mounted on sheets.

Pressing: The specimens should be placed in the field press at the first opportunity, either directly after collection, or sometimes after temporary storage in a vasculum or a polythene bag. A specimen shorter than 15 inches (38 cm) should be kept directly in the folded news paper after loosely spreading the leaves and branches. Herbs, which are generally collected along with the roots, if longer than 15 inches, can be folded in the form of a V, N or W (Fig. 4.4, a-c), always ensuring that the terminal part of the plant with leaves, flowers and fruits is erect, and when finally mounted, the specimens can be easily studied, without having to invert the herbarium sheet. Specimens of grasses and some other groups, which show considerable elasticity, are difficult to hold in a folded condition. These specimens can be managed by using **flexostat** (a strip of stiff paper or card with 2.5 cm long slit). One flexostat inserted at each corner (Fig. 4.4d) holds the specimen in place.

Fig. 4.4   a-c, different methods of folding longer herbaceous plants; d, use of flexostat slips for holding plants in folded condition. Note that the tip of the plant (arrow) would always be erect for convenient study of this important portion with leaves and flowers.

To press bulky fruits, these may be thinly sliced. Large leaves can be trimmed to retain any lateral half. It is useful to invert some leaves so that the undersurface of the leaves can also be studied from a pressed leaf.

Collection and Pressing of Special Groups: A few groups of plants such as conifers, water plants, succulents and mucilaginous plants pose problems during collection and need special methods.

Conifers, although easy to collect and press, pose problems during drying. The tissues of conifers remain living for a long time and progressive desiccation during pressing and drying initiates an abscission layer at the base of leaves and sporophylls. As such a dry twig readily disintegrates, losing its leaves with a slight touch, a problem occasionally encountered in *Abies, Picea, Cedrus* and several other genera. Before pressing, such twigs should be immersed in boiling water for one minute, a pretreatment that kills tissues and prevents the abscission formation during drying. Page (1979) has suggested a pretreatment method involving immersion in 70% ethyl alcohol for 10 min. followed by immersion in 50% aqueous glycerine solution for four days. Since the pretreatment removes the bloom and waxes, and results in a slight colour change, an untreated portion of the plant should be preserved, kept in a small pouch and attached to the herbarium sheet along with the pretreated specimen for reference.

Water plants, especially with submerged leaves, readily collapse due to the absence of cuticle and are difficult to press normally. Such specimens are collected in bags and made to float in a tray filled with water, at the bottom of which a white sheet of paper is placed. The paper is lifted gently, carrying the specimen along, placed in a blotter and pressed. As the slender water plant sticks to the paper, the sheet along with the specimen is shifted from one blotter to another during the process of drying, and finally pasted on the herbarium sheet as such.

Succulents and cacti have a large amount of proliferated parenchyma storing water and, unless special care is taken, these plants readily rot and fungal infection sets in. Such plants are handled by giving slits on thick organs and scooping out the succulent tissue or, alternately, salt is sprinkled on slits to drive out the moisture. The plants may also be killed by pretreatment with ethyl alcohol or formaldehyde.

Mucilaginous plants such as members of the family Malvaceae stick to the blotters and are difficult to process. These plants should be placed between waxed or tissue paper or else folds of muslin cloth. Only the blotter should be changed every time and the specimen separated from the tissue paper or muslin when fully dry.

Aroids and bulbous plants continue to grow even in a press and should be killed with ethyl alcohol and formaldehyde before pressing.

Drying

Drying of pressed plant specimens is a slow process if no artificial heat is used.

Natural drying: A slow process, which may take up to one month for complete desiccation. The plants, freshly collected are placed in a press without corrugated sheets and the press locked for 24 hours. During this **sweating period**, plants lose some moisture, become flaccid and can be easily rearranged. The folded sheet containing the specimen is lifted and placed in a fresh dry folded blotter. In countries using thick coarse newsprint, changing the newspaper is also necessary, and the plant should be carefully transferred from one newspaper to another. The use of a blotter in such a case can be dispensed with, especially after one or two changes. The change of blotters or newspaper sheets is repeated every few days, increasing the interval between the change successively until the specimens are fully dry. The whole process of drying may take about 10 days to one month, depending on the specimens and the climate of the area.

Drying with artificial heat: Drying with the help of artificial heat takes 12 hours to two days. The specimens after the initial sweating period in the field press are transferred to a drying press, with an ample number of corrugated sheets, usually one alternating every folded blotter containing one specimen. The press is kept in a **drier**, a cabinet in which a kerosene lamp or electric bulb warms the air, which dries the specimens by movement through the corrugates. Use of a hot air blower in the cabinet speeds up circulation of the hot air and consequently faster drying is achieved. Sinnott (1983) developed a **solar-powered drier** capable of drying 100 specimens on a sunny day, and attaining a temperature of up to 60° Celsius in the centre of the press. The unit consists of a flat plate collector and a drying box to hold the press. The collector is composed of a wooden frame, a blackened aluminium absorber plate, insulation and a glass or Plexiglas glazing to retain and channel heat into the drying box. One-inch space is provided between the glazing and the absorber plate. The air enters the collector at the open bottom of the collector panel, is heated by conduction from the absorber, rises by convection into the drying box, moves through the corrugates and finally exits from the uncovered top of the drying box, taking with it moisture from the plant specimens. Drying is accomplished in a single day, occasionally two days for complete drying. This solar drier, with practically no operational cost, should provide a right step towards energy conservation.

The rapid drying of specimens using artificial heat has, however, inherent limitations of rendering plants brittle, loss of bloom and some colour change in leaves.

In arid regions the plants can be dried partially during travel by placing the press horizontally on the luggage rack of the vehicle, with the corrugation ducts facing front, forcing the dry wind through the corrugates as the vehicle moves forward.

Specimens pressed and dried are next mounted on herbarium sheets and properly labelled before these can be incorporated in a herbarium.

II. HERBARIUM METHODS

A herbarium is a collection of pressed and dried plant specimens, mounted on sheets bearing a label, arranged according to a sequence and available for reference or study. In practice herbarium is a name given to a place owned by an Institution, which maintains this orderly collection of plant specimens. Most of the well-known herbaria of the world made their beginning from **botanical gardens**.

Botanical Gardens

Although gardens existed in ancient China, India, Egypt and Mesopotamia, these gardens were not botanical gardens in the true sense. They existed for growing food plants, herbs and ornamentals for aesthetic, religious and status reasons. The famous 'hanging gardens' of Babylon in Mesopotamia is a typical example. The first garden for the purpose of science and education was maintained by Theophrastus in his Lyceum at Athens, probably bequeathed to him by his teacher, Aristotle. Credit for establishment of the first modern botanical garden belongs to Luca Ghini (ca 1490-1556), a professor of botany who developed it at Pisa in Italy in 1544. These were followed by botanical gardens at Padua and Florence in 1545.

Botanical gardens have been instrumental in motivating several well-known authors to develop their own **systems** of classification while trying to fit the plants grown in the garden into some previous system of classification—Linnaeus while working at Uppsala and Bernard de Jussieu at Versailles. Although the majority of the botanical gardens house plant species which the climate of the area can support, several well-known botanical gardens have controlled enclosures to support specific plants. Tropical gardens often need indoor growing space—**screen houses** for most plants and **glasshouses** for the majority of cacti and succulents in wet tropical and temperate gardens. Glasshouses in temperate gardens often require winter heating. Botanical gardens play the following important roles:

1. **Aesthetic appeal:** Botanical gardens have an aesthetic appeal and attract a large number of visitors for observation of general plant diversity as also the curious plants, as for example the Great Banyan Tree (*Ficus benghalensis*) in the Indian Botanical Garden at Calcutta.
2. **Material for botanical research:** Botanical gardens generally have a wide range of species growing together and offer ready material for botanical research, which can go a long way in understanding taxonomic affinities.
3. **On-site teaching:** Collection of plants is often displayed according to families, genera or habitats, and can be used for self-instruction or demonstration purposes.
4. **Integrated research projects:** Botanical gardens with rich living material can support broad-based research projects which can integrate

information from such diverse fields as anatomy, embryology, phytochemistry, cytology, physiology and ecology.

5. **Conservation:** Botanical gardens are now gaining increased importance for their role in conserving genetic diversity, as also in conserving rare and endangered species. The proceedings of the Symposium on Threatened and Endangered Species, sponsored by the New York Botanical Garden in 1976, published as *Extinction is Forever*, and the conference on the practical role of botanical gardens in conservation of rare and threatened species sponsored by the Royal Botanical Gardens, Kew and published as *Survival and Extinction*, are among the major examples of the role of botanical gardens in conservation.

6. **Seed exchange:** More than 500 botanical gardens of the world operate an informal seed exchange scheme, offering annual lists of available species and a free exchange of seeds.

7. **Herbarium and library:** Several major botanical gardens of the world have herbaria and libraries as an integral part of their facilities, and offer taxonomic material for research at a single place.

8. **Public services:** Botanical gardens provide information to the general public on identification of native and exotic species, methods of propagation and also supply plant material through sale or exchange.

Major Botanical Gardens   Thousands of botanical gardens located worldwide are maintained by various Institutes. Of these, nearly 800 important gardens are documented in the *International Directory of Botanical Gardens* published by Henderson (1983). A botanical garden today is an area set aside and maintained by an organisation for growing various groups of plants for study, aesthetics, conservation, economic, educational, recreational and scientific purposes. Some of the major botanical gardens are listed below:

New York Botanical Garden, USA (Fig. 4.5)—Christened the New York Botanical Garden in 1891, when the Torrey Botanical Club adopted its

Fig. 4.5   Haupt conservatory complex of New York Botanical Garden.

foundation as a corporation chartered by the State. David Hosak founded the garden in 1801 as Algin Botanic Garden. Professor N.L. Britton, the most productive taxonomist of his time, directed the idea of advancement of botanical knowledge through research at this botanical garden. The garden today covers 100 ha in the heart of New York City along the Bronx River. In addition 778 ha Mary Flager Cary Arboretum at Millbrook have been added to the jurisdiction of the garden. There are 15,000 species distributed in the demonstration gardens, Montgomery conifer collection, Stout day lily garden, Havemeyer lilac collection, *Rhododendron* and *Azalea* collection, Everett rock garden, herb garden, rose garden, arboretum and conservatory complex. The garden has a systematic arrangement of trees and shrubs that make it a place of interest for the general public as well as botanists and horticulturists. The garden plays a major role in conservation of rare and endangered species.

The garden has a well-maintained herbarium of over 5 million specimens from all over the world, but mainly from the New World. The library houses over 200,000 volumes and over 500,000 items (including pamphlets, photographs, letters etc.). It also maintains a huge botanical database.

Royal Botanic Gardens, Kew (Fig. 4.6)—More popularly known as 'Kew Gardens', is undoubtedly the finest botanical garden and botanical research and resource centre in the world. The garden was developed in the 1600s by Kew House owned by Richard Bennet. The widow of the Prince of Wales commissioned the garden in 1759 and William Aiton took over as its superintendent. Sir Joseph Banks introduced large collections from different parts of the world. In 1841 the management of the garden was transferred from the crown to the parliament and Sir William Hooker became its first official director. He was mainly responsible for the advancement of the garden, enlarging it from a mere 6 ha to more than 100 ha and building a palm house. Sir J.D. Hooker, who succeeded his father as its Director,

Fig. 4.6   Princess of Wales House at Royal Botanic Gardens, Kew.

added rhododendrons, and also authored several important publications. John Hutchinson worked and developed his famous system of classification here.

The garden has since grown into a premier Research and Educational Institute with excellent herbarium and library. Originally the garden covered an area of 120 ha. The outstation of the Royal Botanic Gardens, Kew at Wakehurst Place near Ardingly in West Sussex is a rural estate of 202 ha with an Elizabethan mansion, and was acquired in 1965. The Royal Botanic Gardens, Kew has directed and financed development so that Wakehurst Place now makes a vital contribution in maintaining the international reputation of the Living Collections Department (LCD). In particular the practical in-situ conservation policies pursued, and the rich and diverse plant collections, which are maintained, add greatly to the LCD's activities. The environmental conditions of the High Weald of Sussex contrast with those of Kew by offering varied topography, higher rainfall and more diverse and moisture retentive soils. These combine to provide a range of microclimates, which make possible the successful cultivation of a great diversity of plants, many of which do not thrive at Kew. There are substantive differences in the layout and content of the collections at Wakehurst Place which act to complement those at Kew. In particular the botanical collections are laid out in a floristic manner reflecting the way that temperate plant communities have evolved. The botanical collections are supported by extensive ornamental displays exploiting the wide range of available biotopes and acting as primary visitor attractants. A final element of the woodland cover is forestry plots comprising high forest and Christmas tree plantations. Jodrell Laboratory at Kew has established itself as the world centre in the study of plant anatomy, cytogenetics and plant biochemistry.

The Royal Botanic Gardens' Living collections at Kew and Wakehurst Place are a multilevel encyclopaedic reference collection reflecting global plant diversity and providing a reference source which serves all aspects of botanical and horticultural science within Kew, Great Britain and throughout the world. It is probably the largest and most diverse living collection in the world. The two sites provide quite different environments, allowing the development of two differing but complementary collections. The living collections at Kew are most diverse with 351 families, 5465 genera and over 28,000 species growing successfully. The arboretum covers the greatest area with large mature temperate trees. Tropical plants are maintained indoors, including Aroid House, Palm House, Filmy Fern House etc. Several interesting plants such as *Victoria amazonica* from South America and *Welwitschia mirabilis* from Angola are also growing here.

Kew Herbarium, undoubtedly the most famous herbarium of the world, maintains over 6 million specimens of vascular plants and fungi from every country in the world. There are over 275,000 type specimens as well. The library at Kew is very extensive with over 750,000 books and journals a resource for all Kew's research work. *Kew Bulletin* and *Index Kewensis* are its two continuing premier publications. Kew maintains databases on plant names, taxonomic literature, economic botany, plants for arid lands and on plant groups of special economic and conservation value. Kew also makes about 10,000 identifications a year through its Herbarium service and provides specialist advice on taxonomy and nomenclature in difficult cases. Kew is involved in major biodiversity research programmes in many parts of the world, including tropical and West Asia, SE Asia, Africa, Madagascar, South America, and the Pacific and Indian Oceanic islands. Collaborative research programmes are based in over 40 countries with a concentration in Brunei, Malaysia, Indonesia, East Africa, Cameroon, Madagascar, China and Brazil. The Herbarium runs an international Diploma Course in Herbarium Techniques. The General Catalogue now contains over 122,000 records and is available throughout RBG Kew on the network.

Pisa Botanical Garden, Italy—Pisa Garden, developed by Luca Ghini in 1544, is credited as the first modern botanical garden. The garden was known for the finest specimens of *Aesculus hippocastanum*, *Magnolia grandiflora*, and several other species. Though the garden does not exist today, the records of its design demonstrate geometric outlay of plantings that are characteristic of several continental gardens even today.

Padua Botanical Garden, Italy—The garden is a contemporary of Pisa Botanical Garden, established in 1545. The speciality of this garden is the elegance and Halian taste, which has been wedded to the service of science. The garden elegance and beauty are equalled by Kew Gardens only.

Berlin Botanic Garden and Museum, Berlin-Dahlem—The garden had its beginning in 1679 when the Grand Duke of Berlin gave instructions to open an agricultural model garden in Schoneberg, a village near Berlin. Due to lack of space it was later relocated to Dahlem. The garden developed largely due to the efforts of C.L. Wildenow who built it from an old rundown royal garden. Adolph Engler and L. Diels who were its subsequent directors improved its quality and content. Much of the garden was destroyed during World War II. It was rebuilt largely through the efforts of Robert Pilger, its then director.

The botanical garden today comprises an area of 126 acres. About 20,000 different species of plants are cultivated here. The section on plant geography covering 39 acres—one of the biggest of its kind in the world—depicts the whole of the Northern Hemisphere. The arboretum and taxonomy section cover 42 acres and include around 1800 species of trees and shrubs and nearly 1000 species of herbaceous plants, the latter arranged according to the classification system of Adolph Engler. The botanical museum specialises in the display of botanical exhibits, being the only museum of its kind in Central Europe with models of various life-forms. The main tropical greenhouse (Fig. 4.7) with its length of 60 m and height of 23 m is one of the largest in the world. Featuring tall trees with epiphytes, rich ground vegetation and lianas give an idea of the vast variety of tropical vegetation.

Fig. 4.7   Tropical greenhouse of Berlin Botanic Garden at Dahlem.

Cambridge University Botanical Garden—The garden was founded in 1762 as a small garden on 5 acres of land in the centre of Cambridge. It was moved to the present location in 1831 when Prof. J.S. Henslow established it on newly acquired land of the University covering 40 acres. The garden is artistically landscaped with systematic plantings, winter-hardy trails, an alpine garden and a chronological bed. The latter is in the form of a narrow bed (300 × 7 feet) divided into 24 sections with each containing plants introduced during a 20-year period. Tropical houses are one of the major attractions of the garden and contain palms and other tropical plants.

Herbaria

It was again Luca Ghini who initiated the art of herbarium making by pressing and sewing specimens on sheets of paper. This art was disseminated throughout Europe by his students who mounted sheets and bound them into book volumes.

Although the herbarium technique was a well-known botanical practice at the time of Linnaeus, he departed from the convention of mounting and binding the specimens into volumes. He mounted specimens on single sheets, storing them horizontally, a practice followed even today. From isolated personal collections, herbaria have grown into large institutions of national and international stature with millions of specimens from different parts of the world. *Index Herbariorum*, edited by Patricia Holmgren (Fig. 4.8) (Holmgren, Holmgren and Barnett, 1990) lists the world's important herbaria. Each herbarium is identified by an abbreviation that is valuable in locating the type specimens of various species. The major herbaria of the world with approximate number of specimens in the order of importance are listed in Table 4.1.

Fig. 4.8 Patricia Holmgren of New York Botanical Garden, the editor of *Index Herbariorum* and 2 volumes of *Intermountain Flora* (courtesy New York Botanical Garden, Bronx).

Table 4.1: Major herbaria of the world, listed in the order of number of specimens

	Herbarium	Abbr.	No. of specimens
1.	Museum National d'Histoire Naturelle (Museum of Natural History), Paris	P	10,000,000
2.	Royal Botanic Gardens, Kew	K	6,000,000
3.	Komarov Botanic Institute, St. Petersburg (Leningrad)	LE	5,700,000
4.	Conservatoire et Jardin Botaniques (Conservatory and Botanical Garden), Geneva	G	5,000,000
5.	Combined Herbaria, Harvard University, Cambridge, Massachusetts*	A, FH, GH, AMES	5,000,000
6.	New York Botanical Garden, New York	NY	5,000,000
7.	US National Herbarium (Smithsonian), Washington	US	4,100,000
8.	British Museum of Natural History, London	BM	4,000,000
9.	Institute de Botanique, Montpellier	MPU	4,000,000
10.	Naturhistorika Riksmuseet, Stockholm	S	4,000,000

*Consists of Arnold Arboretum (A), Farlow Herbarium (FH), Gray's Herbarium (GH) and Oaks Ames Orchid Herbarium (AMES).

In India, the Central National Herbarium of the Indian Botanic Garden, Botanical Survey of India, Calcutta (CAL) has over 1.3 million specimens. The herbarium of the Forest Research Institute, Dehra Dun (DD) and the National Botanical Research Institute, Lucknow (LUCK) are other major herbaria in India, with collections from allover the world.

Roles of a Herbarium From a safe place for storing pressed specimens, especially type material, herbaria have gone a long way in becoming major centres of taxonomic research. Additionally, herbaria also form an important link for research in other fields of study. Classification of the world flora is primarily based on herbarium material and associated literature. More recently, herbaria have gained importance as sources of information on endangered species and are of primary interest to conservation groups. The major roles played by a herbarium include:

1. **Repository of plant specimens:** Primary role of a herbarium is to store dried plant specimens, safeguard these against loss and destruction by insects, and make them available for study.

2. **Safe custody of type specimens:** Type specimens are the principal proof of the existence of a species or an infraspecific taxon. These are kept in safe custody, often in rooms with restricted access, in several major herbaria.

3. **Compilation of Floras, Manuals and Monographs:** Herbarium specimens are the 'original documents' upon which the knowledge of taxonomy, evolution and plant distribution rests. Floras, manuals and monographs are largely based on herbarium resources.

4. **Training in herbarium methods:** Many herbaria carry facilities for training graduates and undergraduates in herbarium practices, organising field trips and even expeditions to remote areas.

5. **Identification of specimens:** The majority of herbaria have a wide-ranging collection of specimens and offer facilities for on-site identification or having the specimens sent to the herbarium identified by experts. Researchers can personally identify their collection by comparison with the duly identified herbarium specimens.

6. **Information on geographical distribution:** Major herbaria have collections from different parts of the world and thus scrutiny of the specimens can provide information on geographical distribution of a taxon.

7. **Preservation of voucher specimens:** Voucher specimens preserved in various herbaria provide an index of specimens on which a chromosomal, phytochemical, ultrastructural, micromorphological or any specialised study has been undertaken. In the case of a contradictory or doubtful report, the voucher specimens can be critically examined to arrive at a more satisfactory conclusion.

Mounting of Specimens

Pressed and dried specimens are finally mounted on herbarium sheets. A standard herbarium sheet is **29 by 41.5 cm** ($11\frac{1}{2}$ by $16\frac{1}{2}$ **inches**), made of thick handmade paper or a card sheet. The sheet should be relatively stiff to prevent damage during handling of specimens. It should have a high rag content (preferable 100 per cent) with fibres running lengthwise.

The specimens are attached to the sheet in a number of ways. Many older specimens in the herbaria are frequently found to have been sewn on the sheets. Use of adhesive linen, paper or cellophane strips is an easier and faster method of fixing specimens. Most of the contemporary specimens are fixed using liquid paste or glue in one of two ways, however:

(i) Paste or glue is applied to the backside (if distinguishable) of the specimen, which is later pressed onto the mounting sheet and allowed to dry in pressed condition for a few hours. This method is slower but more economical.

(ii) Paste or glue is smeared on a glass or plastic sheet, the specimen placed on the sheet and the glued specimen transferred to a mounting sheet. This method is more efficient but expensive.

The stem and bulky parts may often require adhesive strips or even sewing for secure fixing of specimens. Small paper envelops called **fragment packets** are often attached to the herbarium sheet to hold seeds, extra flowers or loose plant parts.

Labelling An **herbarium label** is an essential part of a permanent plant specimen. It primarily contains the information recorded in the **field diary** at the time of collection, as also the results of any subsequent identification process. The label is located on the lower right corner of the herbarium sheet (Fig. 4.9), with information recorded on the preprinted proforma, printed directly on the sheet or on the paper slips which are pasted on the sheets. It is ideal to type the information. If handwritten, it should be in permanent ink. Ball pens should never be used as the ink often spreads after some years.

There is no agreement as to the size of a herbarium label, the recommendations being as diverse as $2\frac{3}{4}$ by $4\frac{1}{4}$ inches (Jones and Luchsinger, 1986) and **4 by 6** inches (Woodland, 1991). The information commonly recorded on the herbarium label includes:

(i) Name of the Institution
(ii) Scientific name
(iii) Common or vernacular name
(iv) Family
(v) Locality
(vi) Date of collection
(vii) Collection number
(viii) Name of the collector
(ix) Habit and habitat including field notes.

An expert visiting a herbarium may want to correct an identification or record a name change. Such correction is never done on the original label but on a small **annotation label** or **determination label**, usually **2 by 11 cm** and appended left of the original label. Such a label, in addition to the correction, records the name of the person and the date on which the change was recorded. Such information is useful, especially when more than one annotation label is appended by different persons. The last label is likely to be the correct one.

Voucher herbarium specimens of a research study often have authentic information about the specimens recorded in the form of a **voucher label**.

Fig. 4.9   A sample herbarium sheet with mounted specimen and a label.

Filing of Specimens   Mounted, labelled and treated (to kill insect pests) specimens are finally incorporated in a herbarium, where they are properly stored and looked after. Small herbaria arrange specimens alphabetically according to family, genus and species. Larger herbaria, however, follow a particular system of classification. Most herbaria usually follow Bentham and Hooker (British herbaria and most commonwealth countries) or Engler and Prantl (Europe and North America). Many herbaria of the latter category follow the **number code** of families and genera given by Dalla Torre and Harms (1900-1907).

The specimens belonging to a species are placed in a folder made of thin strong paper, termed **species cover**. The species covers belonging to a particular genus are often arranged alphabetically and placed inside a **genus cover** made of a thicker paper. More than one genus cover may be used if the number of species are more, or if the specimens are to be arranged geographically. The genus covers of a family are arranged according to the system of classification being followed. The demarcation between the two families (last genus of a family and first genus of the next family) is done using a sheet of paper with a **front-hanging label** indicating the name of the next family. The folders are stacked in pigeonholes of the herbarium cases and the arrangement is suitable for shifting of folders as the number of specimens increase with time. Unknown specimens are kept in separate folders marked **dubia**, placed towards the end of a genus (when the genus is identified) or a family (when the family is identified but not the genus), as the case may be, so that an expert can examine them conveniently. Standard **herbarium cases** are insect- and dustproof with two tiers of pigeonholes, each 19 in. deep, 13 in. wide and 8 in. high (Fig. 4.10).

Fig. 4.10 Herbarium cabinet with filed specimens used in New York Botanical Garden Herbarium (photograph courtesy New York Botanical Garden).

Type specimens are usually kept separately in distinct folders or often in separate herbarium cases, sometimes even separate rooms, for better care and safety.

A herbarium commonly maintains an **index register** in which all the genera in the herbarium are listed alphabetically and against each genus is indicated the family number and the genus number, the two helping for convenient incorporation and retrieval of specimens in a herbarium.

Pest Control

Herbarium specimens are generally sufficiently dry and as such not attacked by bacteria or fungi. They are, however, easily attacked by pests such as **silverfish, dermestid beetles** (cigarette beetle, drugstore beetle and black carpel beetle). Control measures include:

1. **Treating incoming specimens:** Specimens have to be pest free before they can be incorporated into a herbarium. This is achieved in three ways:
    (i) *Heating* at temperatures up to 60° C for 4-8 h in a heating cabinet. The method is effective but the specimens become brittle.
    (ii) *Deep-freezers* have now replaced heating cabinets in most herbaria of the world. A temperature of minus 20 to minus 60° C is maintained in most herbaria.
    (iii) *Microwave ovens* have been used by some herbaria, but as indicated by Hill (1983), the use of microwave ovens has some serious shortcomings including:
        a) Stems containing moisture burst due to sudden vaporisation. of water inside.
        b) Metal clip staples on the sheets get overheated and may char the sheet.

c) The embryo in the seed gets killed, thus destroying a valuable source of experimental research, as seeds from herbarium specimens are often used for growing new plants for research projects.

2. **Use of repellents:** Chemicals with an offensive odour or taste are kept in herbarium cases to keep pests away from specimens. **Naphthalene** and **Paradichlorobenzene** (PDB) are commonly used repellents, usually powdered and put in small muslin bags kept in pigeonholes. PDB is more toxic and as such prolonged exposure of workers should be avoided. For people working 8 hours a day in a 5-day per week schedule, the upper exposure level for naphthalene is 75 PPM and for PDB 10 PPM.

3. **Fumigation:** In spite of pretreatment of specimens and use of repellents, the use of fumigants is necessary for proper herbarium management. Fumigation involves exposing specimens to the vapours of certain volatile substances. A mixture of **ethylene dichloride** (3 parts) and **carbon tetrachloride** (1 part) was once commonly used for fumigation. Ethylene dichloride is explosive without carbon tetrachloride, but the latter is extremely toxic to humans, causing liver damage, and as such the use of this fumigant has been banned. **Dowfume-75** has been cleared by the Environmental Protection Agency for use in herbaria. Under controlled conditions **Vapona resin strips** (Raid strips) are suitable for herbarium cases. One-third of a strip is placed in each herbarium case for seven to ten days twice a year. The cases should not be opened during the period of fumigation.

III. IDENTIFICATION METHODS

Identification of an unknown specimen is a common taxonomic activity, and often combined with determination of a correct name. The combined activity is appropriately referred as **specimen determination**. The identification of an unknown plant may be achieved by comparison with identified herbarium specimens or through the help of taxonomic literature. Both methods may be combined for a more reliable identification.

The unknown specimen meant for identification is sent to a herbarium, where an expert on the plant group examines and identifies it by comparison with duly identified specimens. The user can also visit a herbarium and personally compare and identify his specimens.

Computers have entered in a big way into solving identification problems. Electronic revolution in recent years has opened up a new, faster and more reliable method of identification. The photograph, description or illustration of parts can be put up on a web site, with information to a relevant **e-mail list**, whose members can help in achieving identification within hours.

Taxonomic Literature

Various forms of literature incorporating description, illustrations and identification keys are useful for proper identification of unknown plants. The library is therefore as important in taxonomic work as a herbarium and a knowledge of taxonomic literature is vital to the practising taxonomist. The literature of taxonomy is one of the oldest and most complicated literatures of science. Several **bibliographic references, indexes** and **guides** are available to help taxonomists to locate relevant literature concerning a taxonomic group or a geographical region. The major forms of literature helpful in identification are described below.

Floras   A **Flora** is an inventory of the plants of a defined geographical region. A Flora may be fairly exhaustive or simply synoptic. Lists of the Floras may be found in the *Geographical Guide to the Floras of the World* by S.F. Blake (Part I, 1941; Part II, 1961) and *Guide to the Standard Floras of the World* by Frodin (1984). Depending on the scope and the area covered the Floras are categorised as:

1. **Local Flora** covers a limited geographical area, usually a state, county, city, a valley or a small mountain range. Examples: *Flora of Delhi* by J.K. Maheshwari (1963), *Flora Simlensis* by H. Collet (1921), *Flora of Tamil Nadu* by K. M. Mathew (1983), and *Flora of Central Texas* by R.G. Reeves (1972).

2. **Regional Flora** includes a larger geographical area, usually a large country or a botanical region. Examples: *Flora of British India* by Sir J.D. Hooker (1872-97), *Flora Malesiana* by C.G. Steenis (1948), *Flora Iranica* by K.H. Rechinger (1963), and *Flora SSSR* by V.L. Komarov and B.K. Shishkin (1934-64). A Flora covering a country is more appropriately known as a **National Flora**.

3. **Continental Flora** covers the entire continent. Examples: *Flora Europaea* by T.G. Tutin et al. (1964-80) and *Flora Australiensis* by G. Bentham (1863-78).

4. **Comprehensive Treatments** have a much broader scope. Although no World Flora has ever been written, several important works have attempted a worldwide view. Examples: *Genera plantarum* of G. Bentham and J.D. Hooker (1862-83), *Die Naturlichen pflanzenfamilien* of A. Engler and K.A. Prantl (1887-1915) and *Das Pflanzenreich* of A. Engler (1900-54).

Manuals   A **manual** is a more exhaustive treatment than a Flora, always having keys for identification, description and glossary but generally covering specialised groups of plants. Examples: *Manual of Cultivated Plants* by L.H. Bailey (1949), *Manual of Cultivated Trees and Shrubs Hardy in North America* by A. Rehder (1940) and *Manual of Aquatic Plants* by N.C. Fassett (1957).

A manual differs from a monograph in that the latter is a detailed taxonomic treatment of a taxonomic group.

Monographs   A **monograph** is a comprehensive taxonomic treatment of a taxonomic group, generally a genus or a family, providing all taxonomic data relating to that group. Usually the geographical scope is worldwide since it is impossible to discuss a taxon without including all its members, and often all its species, subspecies, varieties and forms are discussed. The monograph also includes an exhaustive review of literature, as also a report on author's research work. A monograph includes all information related to nomenclature, designated types, keys, exhaustive description, full synonymy and citation of specimens examined. Examples: *The Genus Pinus* by N.T. Mirov (1967), *The Genus Crepis* by E.B. Babcock (1947), *A Monograph of the Genus Avena* by B.R. Baum (1977), *The Genus Datura* by A.F. Blakeslee et al. (1959) and *The Genus Iris* by W.R. Dykes (1913).

A **revision** is less comprehensive than a monograph, incorporating less introductory material and including a synoptic literary review. A revision includes a complete synonymy but the descriptions are shorter and often confined to diagnostic characters. The geographical scope is usually worldwide.

A **conspectus** is an effective outline of a revision, listing all the taxa, with all or major synonyms, with or without short diagnosis and with brief mention of the geographical range. *Species plantarum* of C. Linnaeus (1753) is an ideal example.

A **synopsis** is a list of taxa with very abbreviated diagnostic distinguishing statements, often in the form of keys.

Journals   Whereas floras, manuals and monographs are published after a lot of taxonomic input and it may take several decades before they are revised, if at all, taxonomic journals provide information on the results of ongoing research. A continuous update on additional taxa described or reported from a region, nomenclatural changes and other taxonomic information is essential for continuing taxonomic activity. Reference to a publication in a journal includes volume number (all issues within a year bear the same volume number), issue number (numbered within a volume, a monthly journal would have 12 issues, quarterly 4 issues and so on) and page numbers on which a particular article appears. Common journals devoted largely to taxonomic research include: *Taxon* (International Association of Plant Taxonomy, Berlin), *Kew Bulletin* (Royal Botanic Gardens, Kew), *Plant Systematics and Evolution* (Denmark), *Botanical Journal of Linnaean Society* (London), *Journal of the Arnold Arboretum* (Harvard), *Bulletin Botanical Survey of India* (Calcutta), *Botanical Magazine* (Tokyo) and *Systematic Botany* (New York).

Supporting Literature   With a large amount of research material being published throughout the world, there is always need for supporting literature to give consolidated information about the works published world over. They also help in tracking down material concerning a particular taxon covering a certain period. *Taxonomic Literature* an exhaustive series of *Regnum vegetabile*, covers full bibliographical details of literature very helpful in searching type material, priority of names, dates of publication and biographic data on authors. Originally published in 1967, it is under constant revision with 3 supplements of the 2nd edition published between 1992-97 (Stafleu and Mennega).

Abstracts or **Abstracting journals** provide a summary of various articles published in various journals throughout the world. *Biological Abstracts* and *Current Advances in Plant Science* are more general in approach. *Kew Record of Taxonomic Literature* covers all articles relevant to taxonomy.

An **Index** provides an alphabetic listing of taxa with reference to their publication. *Index Kewensis* is by far the most important reference tool, first published in 2 volumes from Royal Botanic Gardens, Kew (1893-1895), covering names of species and genera of seed plants published between 1753 and 1885. Regular *Supplements* used to be published every 5 years and 18 *Supplements* appeared up to 1985. Since then the listing has been published annually under the title *Kew Index*.

Index Kewensis (Fig. 4.11) is a list of new and changed names of seed-bearing plants with bibliographic references to the place of first publication. Supplement 19 was published in 1991 covering the years 1986 to 1990. At the beginning of the nineteen eighties the data were transferred to a computer database which continues to expand at the rate of approximately 6000 records per year. To make this data generally available it was decided to publish the whole *Index Kewensis* as a CD-ROM in 1993. This contains almost 968,000 records.

URTICA, [Tourn.] Linn. Syst. ed. 1(1785). *URTI-*
    CACEAE, Benth. & Hook. f.iii. 381.
        ADICEA, Rafin. Cat. 18 (1824).
        ILDEFONSIA, Mart. ex Steud. Nom. ed. II. i. 802 (1840).
        RUTICA, Neck. Elem. ii. 202 (1790).
        SELEPSION, Rafin. Fl. Tellur. iii. 48 (1836).
    acerifolia, Zenker, Fl. Ind. Dec. i. tt. 3, 4,=Girardinia palmata.
    acuminata, Poir. Encyc. Suppl. iv. 224 = Urera acuminata.
    adoensis, Hochst. in Flora, xxiv. (1841) 1. Intell. 21= Girardinia condensata.

Fig. 4.11 A page from *Index Kewensis*.

Illustrations of vascular plants can be located through *Index Londinensis*, which contains information up to 1935. More recent information can be found in the 2-volume work *Flowering Plant Index of Illustrations and Information* compiled by R.T. Isaacson (1979). A listing of all generic names can be found in *Index nominum genericorum* (ING) a 3-volume work published in 1979 under the series *Regnum Vegetabile*. The first supplement appeared in 1986. It has now been put on the database and can be directly accessed through the Internet. *Index holmiensis* (earlier *Index holmensis*) is an alphabetic listing of distribution maps found in taxonomic literature of vascular plants. It started publication in 1969. *Gray Herbarium Card Index* is information on cards, which has now been set up on a database. Usually on the same pattern as *Index Kewensis*, the Index has been published in 10 volumes between 1893 to 1967. A 2-volume *supplement* was published by G.K. Hall in 1978. The *Gray Herbarium Index* database currently includes 287,225 records (in 1998) of New World vascular plant taxa at the level of species and below. The Index includes from its 1886 starting point, the names of plant genera, species and all taxa of infraspecific rank. The *Gray Index* has in common with *Index Kewensis* its involvement with taxon names, although they differ in biological and geographical coverage. The *Gray Index* covers vascular plants of the Americas, *Index Kewensis* includes seed plants worldwide. Only the *Gray Index* has nomenclatural synonyms cross-referenced to basionyms. The information is now accessible over the Internet via keyword searches from the E-mail Data Server and through the Biodiversity and Biological Collections Gopher. Indices covering other groups of plants have also been published: *Index Filicum* for Pteridophytes, and *Index Muscorum* for Bryophytes.

Numerous valuable **Dictionaries** have been published but by far the most useful is *Dictionary of Flowering Plants and Ferns* published by J.C. Willis. The 8th edition revised by Airy Shaw appeared in 1973. The book contains valuable information concerning genera and families, providing name of the author, distribution, family and number of species in the genus.

Taxonomic Keys

Taxonomic keys are *aids for rapid identification of unknown plants*. They constitute important components of Floras, Manuals, Monographs and other forms of literature meant for the identification of plants. In addition, identification methods in recent years have incorporated the usage of keys based on cards, tables and computer programs. The latter are primarily designed for identification by non-professionals. These keys are primarily based on characters, which are stable and reliable. The keys are helpful in a faster preliminary identification, which can be backed up by confirmation through comparison with the detailed description of the taxon provisionally identified with. Before identification is attempted, however, it is necessary that the unknown plant be carefully studied, described and a list of its characters prepared. Based on the arrangement of characters and their utilisation, two types of identification keys are differentiated:

1. Single-access or sequential keys,
2. Multiaccess or multientry keys (polyclaves).

Single-access or Sequential Keys Such keys are usual components of Floras, Manuals, Monographs and other books meant for identification. The keys are based on **diagnostic** (important and conspicuous) characters (**key characters**) and as such the keys are known as **diagnostic keys**. Most óf the keys in use are based on pairs of contrasting choices and as such are **dichotomous keys**. They were first introduced by J.P. Lamarck in his *Flore Française* in 1778. The construction of a dichotomous key starts with preparation of a list of reliable characters for the taxon for which the key is to be constructed. For each character the two contrasting choices are determined (e.g., habit woody or ɪᴇrbaceous). Each choice constitutes a **lead** and the two contrasting choices form a **couplet**. For characters having more than two available choices the character can be split to make it dichotomous. Thus if flowers in a taxon could be red, yellow or white the first couplet would constitute flowers red vs. non-red and the second couplet flowers yellow vs. white. We shall illustrate the construction of keys taking an example from family Ranunculaceae. The diagnostic characters of some representative genera are listed below:

1. *Ranunculus*—Plants herbaceous, fruit achene, distinct calyx and corolla, spur absent, petal with nectary at base.
2. *Adonis*—Plants herbaceous, fruit achene, calyx and corolla differentiated, spur absent, petals without nectary.
3. *Anemone*—Plants herbaceous, fruit achene, calyx not differentiated, perianth petaloid, spur absent.
4. *Clematis*—Plants woody, fruit achene, calyx not differentiated, perianth petaloid, spur absent.
5. *Caltha*—Plants herbaceous, fruit follicle, calyx not differentiated, perianth petaloid, spur absent.
6. *Delphinium*—Plants herbaceous, fruit follicle, calyx not differentiated, perianth petaloid, spur one in number.
7. *Aquilegia*—Plants herbaceous, fruit follicle, callyx petaloid not differentiated from corolla, spurs five in number.

Based upon the above information the following couplets and leads can be identified:

1. Plants woody
   Plants herbaceous
2. Fruit an achene
   Fruit a follicle
3. Calyx and corolla differentiated
   Calyx and corolla not differentiated
4. Spur present
   Spur absent
5. Number of spurs 1
   Number of spurs 5
6. Petal with nectary at base
   Petal without nectary at base

It must be noted that three choices are available for spur (absent, one, five). It has been broken into two couplets to maintain the dichotomy. Based on the arrangement of couplets and their leads three main types of dichotomous keys are in use: **Yoked** or **Indented key, Bracketed** or **Parallel key,** and **Serial** or **Numbered key.**

1. **Yoked** or **Indented key:** This is one of the most commonly used keys in Floras and Manuals especially when the keys are smaller in size. In this type of key the statements (leads) and the taxa identified from them are arranged in visual groups or yokes and additionally the subordinate couplets are indented below the primary one at a fixed distance from the margin, the distance increasing with each subordinate couplet. We shall select the fruit type as the first couplet as it divides the group into two almost equal halves and the taxa excluded would be almost equal whether the fruit in the unknown plant is an achene or a follicle. The yoked or indented key for the taxa under consideration is shown below:

1. Fruit an achene
    2. Calyx differentiated from corolla
        3. Petal with basal nectary .... 1. *Ranunculus*
        3. Petal without basal nectary ....2. *Adonis*
    2. Calyx not differentiated from corolla
        4. Plants woody.....................................4. *Clematis*
        4. Plants herbaceous .... 3. *Anemone*
1. Fruit a follicle
    5. Spur present
        6. Number of spurs 1 .... 6. *Delphinium*
        6. Number of spurs 5 .... 7. *Aquilegia*
    5. Spur absent... 5 *Caltha*

It is important to note that all genera with achene fruit appear together and form visual groups, leads of subordinate couplets are at increasing distance from the margin and the leads of initial couplets are far separated whereas those of subsequent subordinate couplets are closer. Such an arrangement is very useful in shorter keys especially those appearing on a single page, but if the key is very long running into several pages, an Indented key exhibits important drawbacks. Firstly, it becomes difficult to locate the alternate leads of initial couplets as they may appear on any page. Secondly, with the number of subordinate couplets increasing substantially the key becomes more and more sloping thus reducing the space available for writing leads. This may result in wastage of a substantial page space. The problem is clearly visible in *Flora Europaea* where attempts to reduce the indentation distance in longer keys has further complicated the usage of keys. These two disadvantages are taken care of in the Parallel or Bracketed key.

2. **Bracketed** or **Parallel key:** This type of key has been used in larger floras such as *Flora of the USSR, Plants of Central Asia* and *Flora of the British*

Isles. The two leads of a couplet are always together and the distance from the margin is always the same as illustrated below:

1. Fruit an achene...2
1. Fruit a follicle 5
2. Calyx differentiated from coroll 3
2. Calyx not differentiated from corolla 4
3. Petal with basal nectary...................... 1. *Ranunculus*
3. Petal without basal nectary .... 2. *Adonis*
4. Plants woody .... 4. *Clematis*
4. Plants herbaceous .... 3. *Anemone*
5. Spur present 6
5. Spur absent .... 5. *Caltha*
6. Number of spurs 1 .... 6. *Delphinium*
6. Number of spurs 5 .... 7. *Aquilegia*

Several variations of this are used wherein the second lead of the couplet is not numbered as in *Flora of British Isles* or else the second lead is prefixed with a + sign as in *Plants of Central Asia*. The arrangement of couplets in this type of key is useful for longer keys as location of alternate keys is no problem (two are always together) and there is no wastage of page space. There is, however, one associated drawback, the statements are no longer in visual groups. Retention of positive features of the Parallel key and visual groups of the Yoked key is achieved in the Serial key.

3. **Serial** or **Numbered key:** Such a key has been used for the identification of animals and also adopted in some volumes of *Flora of the USSR*. The key retains the arrangement of the Yoked key, but with no indentation so that distance from the margin remains the same. The location of alternate leads is made possible by serial numbering of couplets (or leads when separated) and indicating the serial number of the alternate lead within parentheses. A serial key for the taxa in question would appear as under:

1.(6) Fruit an achene
2.(4) Calyx differentiated from corolla
3.   Petal with basal nectary................... 1. *Ranunculus*
3.   Petal without basal nectary .... 2. *Adonis*
4.(2) Calyx not differentiated from corolla
5.   Plants woody..................................... 4. *Clematis*
5.   Plants herbaceous .... 3. *Anemone*
6.(1) Fruit a follicle
7.(9) Spur present
8.   Number of spurs 1 .... 6. *Delphinium*
8.   Number of spurs 5 .... 7. *Aquilegia*
9.(7) Spur absent.. 5. *Caltha*

Such a key retains the visual groups of statements and taxa, alternate leads, even though separated, are easily located and the there is no wastage of page space.

An inherent drawback of dichotomous keys is that the user has a single fixed choice of the sequence of characters decided by the person who constructs the key. In the said example if information about the fruit is not available, it is not possible to go beyond the first couplet.

Guidelines for Constructing Dichotomous Keys Certain basic considerations are important for the construction of dichotomous keys. These include:

1. The keys should be strictly dichotomous, consisting of couplets with only two possible choices.
2. The two leads of a couplet should be mutually exclusive, so that the acceptance of one should automatically lead to the rejection of another.
3. The statements of the leads should not be overlapping. Thus the two leads 'leaves 5-25 cm long' and 'leaves 20-40 cm long' would find it difficult to place taxa with leaves that are between 20 and 25 cm in length.
4. The two leads of a couplet should start with the same initial word. In our example both leads of the first couplet start with 'Fruit'.
5. The leads of two successive couplets should not start with the same initial word. In our example the word 'spur' appears in two successive couplets and as such in the second one the language has been changed to start with 'Number'. If such a change were not possible it would be convenient to prefix the second couplet with 'The'. Thus the other alternative for the second couplet would have the two leads worded as 'The spur 1' and 'The spurs 5'.
6. For identification of trees two keys should be constructed based on vegetative and reproductive characters separately. As trees commonly have leaves throughout the major part of the year, and flowers appear briefly when in many trees leaves are not yet developed, such separate keys are essential for identification round the year.
7. Avoid usage of vague statements. Statements such as 'Flowers large' vs. 'Flowers small' may often be confusing during actual identification.
8. An initial couplet should be selected in such a way that it divides the group into more or less equal halves, and the character is easily available for study. Such a selection would make the process of exclusion faster, whichever lead is selected.
9. For dioecious plants it is important to have two keys based on male and female flowers separately.
10. The leads should be prefixed by numbers or letters. This makes location of leads easier. If left blank the location of leads is very difficult especially in longer keys.

The keys described above have a single character included in a couplet, with two contrasting statements about the character in the two leads. Such keys are known as **monothetic sequential keys**. The commonest forms of keys used in floras, however, have at least some couplets (Fig. 4.12) with several statements about different characters in each lead. These keys are known as **polythetic sequential keys**. Such polythetic keys, also known as synoptic keys, are especially useful for constructing keys for higher categories. Such keys have three basic advantages over the monothetic keys:

1. One or more characters may be unobservable due to damage or non-occurrence of requisite stage in the specimen. In such cases a monothetic key becomes useless.
2. User can make a mistake in deciding about a single character. This error gets minimised if more than one character is used.
3. The single character used in the couplet may be exceptional. Such likelihood is not possible when more than one character is used.

1 Stem woody at base; achenes 3.5-5 mm	**8. pustulatus**
1 Stem not woody; achenes 2-3.75 mm	
2 Annual or biennial	
3 Achenes smooth at least between the ribs; strongly compressed and ± winged	**1. asper**
3 Achenes rugose or tuberculate between the ribs, neither strongly compressed nor winged	
4 Leaf-lobes strongly constricted at base, or narrowly linear; terminal lobe usually about as large as lateral lobes; ligules longer than corolla-tube; achenes abruptly contracted at base	**2. tenerrimus**
4 Leaf-lobes (if present) not constricted at base; terminal lobe usually much larger than lateral lobes; ligules about as long as corolla-tube; achenes gradually narrowed at base	**3. oleraceous**
2 Perennial	

Fig. 4.12 Portion of a polythetic key of the yoked type used in *Flora Europaea* for genus *Sonchus* (vol. 4, p. 327).

Multiaccess or Multientry Keys (Polyclaves)

Such multientry order-free keys are user oriented. Many choices of the sequence of characters are available. Eventually it is the user who decides the sequence in which to use the characters, and even if the information about a few characters is not available, the user can go ahead with identification. Interestingly, identification may often be achieved without having to use all the characters available to the user. Such identification methods often make use of cards. Two basic types of cards are in use:

1. Body punched cards: These cards are also named **window cards** or **peek-a-boo cards**, and make use of cards with appropriate holes in the body of the card (Fig. 4.13). The process involves using one card for one attribute (character-state). In our example we shall need 11 cards (we have chosen only diagnostic characters above, whereas our list in polyclaves could include more characters, and thus more cards to make it more flexible).

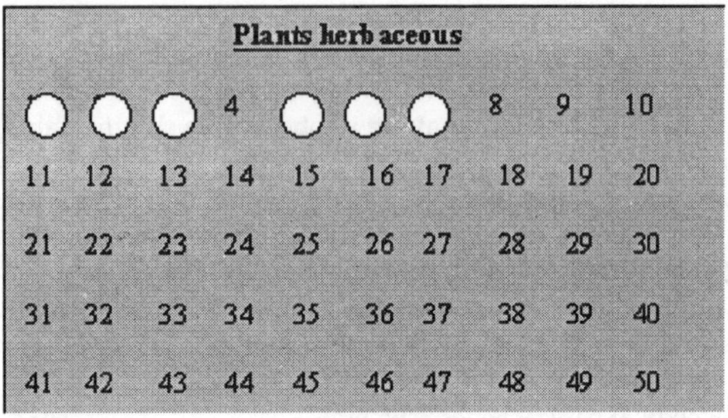

Fig. 4.13 A window card for herbaceous habit for the seven representative genera of Ranunculaceae: 1—*Ranunculus*, 2—*Adonis*, 3—*Anemone*, 4—*Clematis*, 5—*Caltha*, 6 —*Delphinium*, 7—*Aquilegia*.

It should be noted that we selected 12 leads and 6 couplets, with 4 leads for spur. Now we shall need only three actual attributes: 'spur absent', 'spur 1' and 'spurs 5'. Numbers are printed on the cards corresponding to the taxa for which the identification key is meant. In our example we use only 7 of these numbers corresponding to our 7 genera. On each card holes are punched corresponding to the taxa in which that attribute is present. In our example card 'Habit woody' will have only one hole at number 4 (genus *Clematis*), and the card 'Habit herbaceous' will have holes at 1, 2, 3, 5, 6, 7 (all seven except number 4). Once the holes are punched at appropriate positions in all the cards, we are ready for identification. The user studies the unknown plant and makes a list of attributes, according to the sequence he wishes and the characters available to him.

The user starts the identification process by picking up the first card concerning the first attribute in his list of attributes of the unknown plant. He next picks up the second card concerning the second attribute from his list and places it over the first card. This will close some holes of the first card and some of the second card. Only those holes will remain open which correspond to the taxa which contain both the attributes. The third card is subsequently placed over the first two and the process is repeated with additional cards until finally only one hole is visible through the pack of selected cards. The taxon to which this hole corresponds is the identification of the unknown plant.

2. Edge punched cards: An Edge punched card differs from the body punched card in that there is one card for each taxon and holes are punched all along the edge of the card, one for each attribute. In our example here we shall need seven cards, one for each genus. These holes are normally closed along the edge (Fig. 4.14). For each attribute, present in the taxon the hole is clipped out to form an open notch instead of a circular hole along the edge.

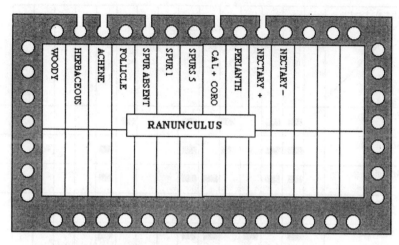

Fig. 4.14 Edge punched card for genus *Ranunculus*. Only the attributes represented in the example above are pictured. Many more attributes could be added along the vacant holes to make the identification process more versatile.

For actual identification all the cards are held together as a pack. A needle is inserted in the hole corresponding to the first attribute of the unknown plant. As this needle is lifted up the taxa containing this attribute would fall down, and those lacking that attribute would remain in the pack lifted by the needle. The latter are rejected. The cards falling down are again arranged in a pack, the needle inserted in the hole corresponding to the next attribute of the unknown plant. The process is repeated until finally a single card falls down. The taxon which this card represents, is the identification of the unknown plant. Note that we may not have to explore all attributes of the unknown plant; identification may be achieved much before we have reached the end of the list of attributes of the unknown plant.

3. Tabular keys: Tabular keys are essentially similar to the polyclaves in that they can take care of exhaustive lists of attributes and are easier to use. The data are incorporated, however, not on cards but in tables with taxa along the rows and attributes along the columns. The attributes represented in each taxon are pictured with the help of appropriate symbols or drawings (Fig. 4.15). The attributes not represented in a taxon show a blank space in the column. Thus the table will have as many rows as taxa and as many columns as number of attributes for which information is available.

The identification process begins with a strip of paper whose width is equal to each row and vertical lines separated by the width of the columns. The attributes present in the unknown plant are pictured on this strip of paper. The strip of paper is next placed towards the top of the table and slowly lowered and compared with each row. The row with which the entries match represents the identification of the unknown plant.

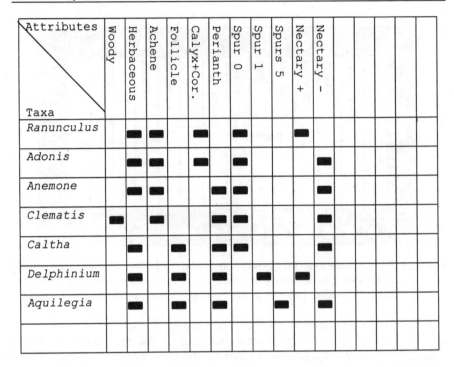

Fig. 4.15 Tabular key for the identification of representative genera of family Ranunculaceae. Only selected attributes as in the above example are pictured. More attributes could be added in additional columns to make the identification process more versatile.

4. Taxonomic formulae: A taxonomic formula is really an alphabetic formula based on a specific combination of alphabets. The various attributes in this method are coded with alphabets. Each taxon thus gets a unique alphabetic formula. These formulae are arranged in alphabetic order in the same manner as words in a dictionary. Based on the attributes of the unknown plant its taxonomic formula is constructed. The next step is as simple as locating a word in the dictionary. The formula is located in the alphabetic list and its identification read against the formula.

The above example of Ranunculaceae could be extended here by assigning alphabets to the attributes: **A:** Woody; **B:** Herbaceous; **C:** Achene; **D:** Follicle; **E:** Spur absent; **F:** Spur 1; **G:** Spurs 5; **H:** Calyx differentiated from corolla; **I:** Calyx not differentiated from corolla, only perianth present; **J:** Nectary present; **K:** Nectary absent. The seven representative genera would thus have the formulae as given below:

ACEIK ....*Clematis*
BCEHJ ....*Ranunculus*
BCEHK ....*Adonis*

BCEIK .... *Anemone*
BDEIK .... *Caltha*
BDFIJ .... *Delphinium*
BDGIK .... *Aquilegia*

Such formulae are really useful in the identification process and have been incorporated in the written version of the multiaccess key to the Genera of Apiaceae in the *Flora of Turkey* (Hedge and Lamond, 1972).

Computers in Identification

Over the years computers have been increasingly used in data collection, processing and integration. They have also found use in a big way in scanning and identifying human ailments, which has greatly helped health management programs. Computers have also found use in plant identification, whereby we no longer need trained botanists for this task. The following main approaches are used in computer identification.

Computer-stored Keys   Dichotomous keys are constructed in the usual manner, fed into a computer and run using an appropriate program, which may be appropriately designed for step-wise processing of the key through a dialogue between the user and the computer. The computer program starts with the first couplet of the key, asks about the attribute in the unknown plant and on the information provided, and handles the key asking relevant questions until finally the actual identification is achieved.

Computer-constructed Keys   Appropriate programs may be developed which can construct a taxonomic key based on the taxonomic information about the taxa, in the same way and based on the same logic which is used by man to construct keys manually. Such keys permanently stored in a computer can be handled as above for the step-wise process of identification.

Simultaneous Character-set Identification Methods   Taxonomic keys are an aid to rapid identification and always provide only a provisional identification, confirmation being achieved only after comparison with a detailed description of the specific taxon. This comparison with the detailed description is not done in the first place as comparing the description of the unknown plant with the description of all taxa of the group or the area would be laborious, time consuming and often impossible. Such a comparison can be achieved through a computer in a matter of seconds. With such an approach the whole set of characters of the unknown plant may be fed into the computer simultaneously and a computer program used to compare the description with the specific group and to suggest the taxon with which the description matches. In case complete information is not available, the computer program may be able to suggest possible alternate identifications.

Automated Pattern-recognition Systems   Computer technology has now developed to the extent that fully automated identification can be achieved. The computer fitted with optical scanners can observe and record features, compare these with those already known and make important conclusions. Programs and techniques are already available for human diagnosis including chemical spectra and photomicrographs of chromosomes, abnormality in human tissues and even in vegetation and agricultural surveys.

DELTA system

The DELTA system is an integrated set of programs based on the DELTA format (DEscription Language for TAxonomy), which is a flexible and powerful method of recording taxonomic descriptions for processing by a computer. DELTA has been adopted as a standard for data exchange by the International Taxonomic Databases Working Group. It enables the generation and typesetting of descriptions and **conventional keys**, conversion of DELTA data for use by **classification programs**, and the construction of **Intkey packages** for interactive identification and **information retrieval**. The System developed in the Natural Resources and Biodiversity Program of the CSIRO Division of Entomology over a period of 20 years, is in use worldwide for diverse kinds of organisms, including fungi, plants, and wood. The programs are continually refined and enhanced in response to feedback from users.

The **DELTA program Key** generates conventional identification keys. Characters are selected by the program for inclusion in the key based on how well the characters divide the remaining taxa. This information is then balanced against subjectively determined weights, which specify the ease of use and reliability of the characters. DELTA data can be readily converted to the forms required by programs for phylogenetic analysis, e.g. Paup, Hennig86 and MacClade. The characters and taxa for these analyses can be selected from the full dataset. Numeric characters are converted into multistate characters, as numeric characters cannot be handled by these programs. Printed descriptions can be generated to facilitate checking of the data.

Identification via e-mail lists

Recent years have seen the emergence of a new efficient method of information exchange through Internet. Illustrations of plants from various parts of the World as also the illustrations of economic plants are put up at various web sites hosted by different Institutions. These illustrations are available for help in identification. A number of **electronic lists** are maintained by list servers. **Taxacom** is one such list very active on taxonomic matters, subscribed to by numerous active taxonomists all over the world. There is a regular exchange on matters of taxonomic interest. Any member with a problem can seek opinions from all members simultaneously. An unknown plant can be identified by sending its description to the list. Still better, a photograph or illustration of the unknown plant can be put up on a website with information to the members. The members may go to the website, observe the photograph or illustration and send their comments to the member concerned or the list itself. The method can work miracles as in the cited example (Fig. 4.16) for which the identification was confirmed within one hour and five minutes.

This innovative method of identification is becoming increasingly popular with more and more members taking help from experts throughout the world.

```
PINE 3.93 MESSAGE TEXT Folder: INBOX Message 91 of 96
Date: Tue, 8 Dec 1998 14:48:16 -0600
From: "Monique D. Reed" <monique@BIO.TAMU.EDU>
To: Multiple recipients of list TAXACOM
<TAXACOM@CMSA.BERKELEY.EDU>
Subject: Mystery plant.

I would be grateful for help in identifying a "mystery"
plant. An image of fruit pieces can be found at http:
//www.csdl.tamu.edu/FLORA/unk.htm
  The bits were collected in the Dominican Republic.
Unfortunately, we do not have seeds or leaves to aid in
identification. We suspect the pieces may be parts of a
capsule of _Hura crepitans_ or a near relative in the
Euphorbiaceae.
Any pointers are welcome.

Monique Reed
Herbarium Botanist
Biology Department
_Texas A&M University
_College Station, TX 77843-3258
_[END of message]

PINE 3.93 MESSAGE TEXT Folder: INBOX Message 93 of 96
ALL
Date: Tue, 8 Dec 1998 15:53:48 -0600
From: "Monique D. Reed" <monique@BIO.TAMU.EDU>
To:   Multiple  recipients  of  list  TAXACOM
<TAXACOM@CMSA.BERKELEY.EDU>
Subject: Mystery solved

So  far,  the  majority  of  the  responses  vote  for  an
identification of Hura crepitans. Many thanks to all who
responded.

Monique Reed
```

Fig. 4.16 Transcript of two e-mail messages to the list Taxacom, first seeking identification of an unknown plant and the second (merely after one hour and 5 minutes) thanking the members for identification.

5

Hierarchical Classifications

It would be total chaos to study and document information about more than a quarter million species of vascular plants if there were no proper mechanism for grouping these plants. Whatever may be the criterion for classification—artificial characters, overall morphology, phylogeny or phenetic relationship—the same basic steps are gone through. The organisms are first recognised and assembled into groups on the basis of certain resemblance. These groups are in turn assembled into larger and more inclusive groups. The process is repeated until finally all the organisms have been assembled into a single, largest most inclusive group. These groups (**Taxonomic groups or Taxa**) are arranged in order of their successive inclusiveness, the least inclusive at the bottom, and the most inclusive at the top.

The groups thus formed and arranged are next assigned to various **categories**, having a fixed sequence of arrangement (**taxonomic hierarchy**), the most inclusive group assigned to the highest category (generally a **division**) and the least inclusive to the lowest category (usually a **species**). The names are assigned to the taxonomic groups in such a way that the name gives an indication of the category to which it is assigned. Rosales, Myrtales, and Malvales all belong to the **order** category and Rosaceae, Myrtaceae and Malvaceae to the **family** category. Once all the groups have been assigned categories and named, the process of classification is complete, or the **taxonomic structure** of the whole largest most inclusive group has been achieved. Because of the hierarchical arrangement of categories to which the groups are assigned, the classification achieved is known as **hierarchical classification**. This concept of categories, groups and taxonomic structure can be illustrated in the form of a **dendrogram** (resembling a pedigree chart—Fig. 5.2 or **box-in-box**—Fig. 5.1).

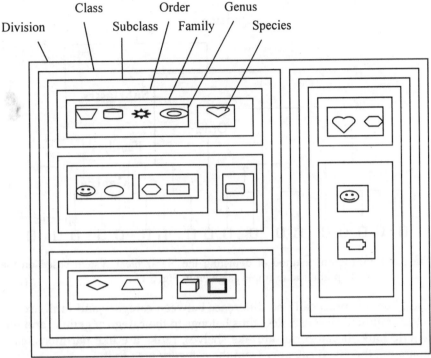

Fig. 5.1 Processes of assembling taxonomic groups according to the hierarchical system, depicted by box-in-box method. In the above example there are 18 species grouped into 10 genera, 6 families, 4 orders, 3 subclasses, 2 classes and 1 division.

I. TAXONOMIC GROUPS, CATEGORIES AND RANKS

The three are inseparable once a hierarchical classification has been achieved. *Rosa alba* is thus nothing else but a **species** and *Rosa* is nothing other than a **genus**. Differences do exist in concept and application, however. The **categories** are like shelves of an almirah, having no significance when empty, and gaining importance and meaning only after something has been placed on them. Thereafter the shelves will be known by their contents: books, toys, clothes, shoes etc. Categories in that sense are artificial and subjective and have no basis in reality. They correspond to nothing in nature. However, they have a fixed position in the hierarchy in relation to other categories. But once a group has been assigned to a particular category the two are inseparable and the category gets a definite meaning because it now includes something actually occurring in nature. The word genus does not carry a specific meaning but the genus *Rosa* says a lot. We are now talking about

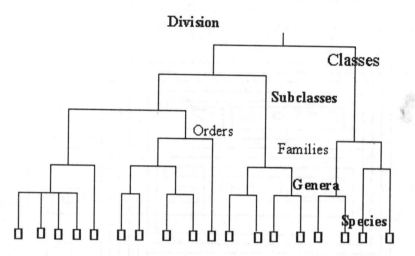

Fig. 5.2 Dendrogram method of depicting the hierarchical system based on the same hypothetical example as given in Fig. 5.1.

roses. There is practically no difference between **category** and **rank**, except in the grammatical sense. *Rosa* thus belongs to the **category genus** , and has **generic rank**. If categories are like shelves, ranks are like partitions, each separating the given category from the category next to it.

Taxonomic groups, on the other hand, are objective and non-arbitrary to the extent that they represent discrete sets of organisms in nature. Groups are biological entities or a collection of such entities. By assigning them to a category and providing an appropriate ending to the name (Rosaceae with ending **-aceae** signifies a family which among others also includes roses, belonging to the genus *Rosa*) we establish the position of taxonomic groups in the hierarchical system of classification. Some important characteristics, which enable a better understanding of the hierarchical system of classification, are enumerated below:

1. Different categories of the hierarchy are higher or lower according to whether they are occupied by more inclusive or less inclusive groups. Higher categories are occupied by more inclusive groups than those occupying lower categories.
2. Plants are not classified into categories but into groups. It is important to note that a plant may be a member of several taxonomic groups, each of which is assigned to a taxonomic category, but is not itself a member of any taxonomic category. A plant collected from the field may be identified as *Poa annua* (assigned to species category). It is a member of *Poa* (assigned to genus category), Poaceae (assigned to family category) and so on, but the plant can't be said to be belonging to the species category.

3. A taxon may belong to other taxa, but it can be a member of only one category. *Urtica dioica* thus is a member of *Urtica*, Urticaceae, Urticales, and so on, but it belongs only to the species category.

4. Categories are not made up of lower categories. The category family is not made up of the genus category, since there is only one genus category.

5. The characters shared by all members of a taxon placed in a lower category provide the characters for the taxon immediately above. Thus the characters shared by all the species of *Brassica* make up the characters of the genus *Brassica*. The characters shared by *Brassica* and several other genera form distinguishing characters of the family Brassicaceae. It is important to note that the higher a group is placed in the hierarchy, the fewer will be the characters shared by the subordinate units. Many higher taxa, as such (e.g. Dicots: Magnoliopsida) can only be separated by a combination of characters; no single diagnostic character may distinguish the taxa. Dicots are thus conveniently separated from monocots by possession of two cotyledons, pentamerous flowers, reticulate venation and vascular bundles in a ring as against one cotyledon, trimerous flowers, parallel venation and scattered vascular bundles in monocots. But when taken individually *Smilax* is a monocot with reticulate venation and *Plantago* is a dicot with parallel venation. Similarly *Nymphaea* is a dicot with scattered bundles, and flowers are trimerous in *Phyllanthus*, which is a dicot.

II. UTILISATION OF CATEGORIES

Taxonomic categories possess only relative value and an empty category has no foundation in reality and obviously can't be defined. An important step in the process of classification is to assign taxa to an appropriate category. It thus becomes imperative to decide what should be the properties for taxa to be included in a particular category? Only with a proper utilisation of the concept of categories can their application in hierarchical systems be meaningful. The problem is far from resolved. An attempt will be made here to discuss the relevant aspects of the inclusion of type of entities or groups of entities under different categories.

Species Concept

Darwin aptly said: 'Every biologist knows approximately what is meant when we talk about species, yet no other taxon has been subjected to such violent controversies as to its definition'. A century and a third has passed, so much advancement in the taxonomic knowledge has been achieved, yet the statement of Darwin is as true today as it was then. Numerous definitions of species have been proposed, making it futile to recount all. Some significant aspects of the problem will be discussed here. Probably the best explanation of diversity of opinions can be explained as under.

The species is a concept. Concepts are constructed by the human mind, and as humans think differently we have so many definitions of a species. Obviously a concept can't have a single acceptable definition.

The word species has a different meaning for different botanists. According to ICBN, which has attempted to clarify the meaning of the word species, 'species are convenient classificatory units defined by trained biologists using all information available'. The word species has a dual cornotation in biological science. First, the species is a naturally occurring group of individual organisms that comprises a basic unit of evolution. Second, the species is a category within a taxonomic hierarchy governed by various rules of nomenclature.

Species as Basic Unit of Taxonomy The following information serves to substantiate the view that species constitutes the basic unit of classification or for that matter taxonomy:

1. Species is considered the basic unit of taxonomy since in the greater majority of cases we do not have infraspecific names. This is especially common in families such as Apiaceae (Umbelliferae) and Liliaceae.

2. Species, unlike other taxa, can be described and recognised without relating to the taxa at other ranks. Thus we can sort herbarium sheets into different species without difficulty, without knowing or bothering to know how many genera are covered by these sheets. We cannot recognise genera or describe them without reference to the included species. Species is thus the only category dealing directly with the plants.

3. Whether defined in terms of morphological discontinuity or restriction of gene exchange, species is unique in being **non-arbitrary to both inclusion and exclusion**. A group is non-arbitrary to inclusion if all its members are continuous by an appropriate criterion. It would be arbitrary to inclusion if it shows internal discontinuity. A group is non-arbitrary to exclusion if it is discontinuous from any other group by the same criterion. A group not showing discontinuity with other groups is arbitrary. All higher taxa although non-arbitrary to exclusion are arbitrary to inclusion, i.e. they show internal discontinuity as now species with external discontinuity form part of these taxa.

Ideal Species A perfect situation! Species that can be easily distinguished and have no problem of identity. Such species, however, are very few; common examples occur in Apiaceae, Asteraceae and the genera *Allium* and *Sedum*. The following characteristics are expected in an ideal species:

1. The species poses no taxonomic problems and is easily recognised as a distinct entity.

2. It exhibits no discontinuity of variation within, i.e., it contains no sub-species, varieties or formas.

3. It is genetically isolated from other species.
4. It reproduces sexually.
5. It is at least partially outbreeding.

Idea of Transmutation of Species This is an ancient Greek idea which persisted as late as the 17th century. Greeks believed in the transmutation of wheat into barley, Crocus into Gladiolus, barley into oats, and many other plants, under certain conditions. It can be explained as nothing other than the result of unintentional mixing of seeds or other propagules of another plant with a particular crop before plantation. The concept is now firmly rejected.

Nominalistic Species Concept This concept is also only of academic interest now. For the purpose of nomenclature all organisms must be referable to species. Species by this concept *can be defined by the language of formal relations and not by property of their organisms*. The concept considers species to be a category in taxonomic hierarchy and may correspond to a specific name in the binomial system of nomenclature. The concept is logically sound but scientifically irrelevant since the ultimate aim is to place a particular group of individuals in a species.

Typological Species Concept This concept was first proposed by John Ray (1686) and further elaborated by C. Linnaeus in *Critica botanica* (1737). Linnaeus refuted the idea of transmutation of species. Linnaeus believed that although there is some variation within a species, the species by themselves are fixed (**fixity of species**) as created by the Almighty Creator. The species according to the concept is *a group of plants which breed true within their limits of variation*. Towards the latter part of his life Linnaeus (*Fundamenta fructificationis*, 1762) imagined that at the time of creation, there arose as many genera as were the individuals. These in the course of time were fertilised by others to give rise to species at a later date. The typological concept, however, should not be confused with **typification**, which is a distinct methodology of Nomenclature, providing names to taxonomic groups.

Taxonomic Species Concept The doctrine of fixity was challenged by Lamarck (1809) and finally Darwin (1859) who recognised continuous and discontinuous variation and developed his taxonomic species concept based on morphology, more appropriately known as the **morphological species concept**. According to this concept the species is regarded as an *assemblage of individuals with morphological features in common, and separable from other such assemblages by correlated morphological discontinuity in a number of features*. The supporters of this view believe in the concept of continuous and discontinuous variations. The individuals of a species show continuous variation, share certain characters and show a distinct discontinuity with individuals belonging to another species, with respect to all or some of these characters.

Du Rietz (1930) modified the taxonomic species concept by also incorporating the role of geographic distribution of populations and developed the *morpho-geographical species concept*. The species was defined as the *smallest population that is permanently separated from other populations by distinct discontinuity in a series of biotypes*.

The populations recognised as distinct species and occurring in separate geographical areas are generally quite stable and remain so even when grown together. There are, however, examples of a few species pairs, which are morphologically quite distinct, well adapted to respective climates, but when grown together they readily interbreed and form intermediate fertile hybrids, bridging the discontinuity gap between the species. Examples are *Platanus orientalis* of the Mediterranean region and *P. occidentalis* of E. United States. Another well-known pair is *Catalpa ovata* of Japan and China and *C. bignonioides* of America. Such pairs of species are known as **vicarious species** and the **phenomenon** as **vicariance** or **vicariism**.

Morphological and **morpho-geographical** types of taxonomic species have been widely accepted by taxonomists who even take into account the data from genetics, cytology, ecology etc. but firmly believe that **species recognised must be delimited by morphological characters.**

The taxonomic species concept has several advantages:

1. It is useful for general taxonomic purposes especially the field and herbarium identification of plants.
2. The concept is very widely applied and most species have been recognised using this concept.
3. The morphological and geographical features used in the application of this concept can be easily observed in populations.
4. Even experimental taxonomists who do not recognise this concept, apply this concept in cryptic form.
5. The greater majority of species recognised through this concept correspond to those established after experimental confirmation.

The concept, however, also has some inherent drawbacks:

1. It is highly subjective and different sets of characters are used in different groups of plants.
2. It requires much experience to practise this concept because only after considerable observation and experience can a taxonomist decide the characters which are reliable in a particular taxonomic group.
3. The concept does not take into account the genetic relationships between plants.

Biological Species Concept

This concept was first developed by Mayr (1942) and defined species as *groups of actually or potentially interbreeding natural populations, which are reproductively isolated from other such groups*. The words 'actually or potentially' being meaningless were subsequently dropped by Mayr (1969). Based on the same criteria Grant (1957) defined species as *a community of cross fertilizing individuals linked together by bonds of mating and reproductively isolated from other species by barriers to mating*. The recognition of biological species thus involves: (a) interbreeding among populations of the same species and (b) reproductive isolation between populations of different species. Valentine and Love (1958) pointed out that species could be defined in terms of gene exchange. *If two populations are capable of exchanging genes freely either under natural or artificial conditions, the two are said to be conspecific (belonging to the same species). On the other hand, if the two populations are not capable of exchanging genes freely and are reproductively isolated, they should be considered specifically distinct*. The concept has several advantages:

1. It is objective and the same criterion is used for all the groups of plants.
2. It has a scientific basis as the populations showing reproductive isolation don't intermix and the morphological differences are maintained even if the species grow in the same area.
3. The concept is based on the analysis of features and does not need experience to put it into practice.

The concept, first developed for animals, holds true because animals as a rule are sexually differentiated and polyploidy is very rare. When applying this concept to plants, however, a number of problems are encountered:

1. A good majority of plants show only vegetative reproduction, and hence the concept of reproductive isolation as such cannot be applied.
2. Reproductive isolation is commonly verified under experimental conditions, usually under cultivation. It may have no relevance for wild populations.
3. Genetic changes causing morphological differentiation and those causing reproductive barriers do not always go hand in hand. *Salvia mellifera* and *S. apiana* are morphologically distinct (two separate species according to the taxonomic species concept) but not reproductively isolated (single species according to the biological species concept). Such species are known as **compilospecies**. Contrary to this, *Gilia inconspicua* and *G. transmontana* are reproductively isolated (two separate species according to the biological species concept) but morphologically similar (single species according to the taxonomic species concept). Such species are known as **sibling species**.
4. Fertility-sterility is only of theoretical value in allopatric populations.
5. It is difficult and time consuming to carry out fertility-sterility tests.
6. Occurrence of reproductive barriers has no meaning in apomicts.
7. Necessary genetic and experimental data are available for only a very few species.

Stebbins (1950) it would appear combined two concepts when he stated that *species must consist of systems of populations that are separated from each other by complete or at least sharp discontinuities in the variation pattern, and that these discontinuities must have a genetic basis*. These populations with isolating mechanisms (different species) may occur in the same region (sympatric species) or in different regions (allopatric species).

Fortunately, although the taxonomic and biological concepts are based upon different principles, the species recognised by one concept, in the majority of cases stand the test of the other. Morphology provides the evidence for putting the genetic definition into practice.

Evolutionary Species Concept This concept was developed by Meglitsch (1954), Simpson (1961) and Wiley (1978). Although maintaining that interbreeding among sexually reproducing individuals is an important component in species cohesion, this concept is compatible with a broad range of reproductive modes. Wiley (1978) defines: *an evolutionary species is a single lineage of ancestor-descendant populations which maintains its identity from other such lineages, and which has its own evolutionary tendencies and historical fate*. This concept avoids many of the problems of the biological concept. Lineage is a single series of demes (populations) that share a common history of descent, not shared by other demes. The identity of species is based on **recognition systems** that operate at various levels. In sexually reproducing species, such systems include recognition because of phenotypic, behavioural and biochemical differences. In asexual species phenotypic, genotypic differences maintain the identity of species. Identity in both sexual and asexual species may also be due to distinct ecological roles.

Several other terms have been proposed to distinguish species based on specific criteria. Grant (1981) recognises **microspecies** as 'populations of predominantly uniparental plant groups which are uniform themselves and are slightly differentiated morphologically from one another'; they are often restricted to a limited geographical area and frequently of hybrid origin. These may be distinguished as **clonal microspecies** (reproducing by vegetative propagation, e.g. *Phragmites*), **agamospermous microspecies** (reproducing by agamospermy, e.g. *Rubus*), **heterogamic microspecies** (reproducing by genetic systems, e.g. *Oenothera biennis* or *Rosa canina*), and **autogamous microspecies** (predominantly autogamous and chromosomally homozygous, e.g. *Erophila*). The term microspecies was first suggested by Jordan (1873) and as such they are often termed **Jardanons** to distinguish them from **Linnaeons**, the normal species, first established by Linnaeus. Microspecies are distinct from **cryptic species**, which are morphologically similar but cytologically or physiologically different. Stace (1989) uses the term **semicryptic species** for the latter.

The term **biosystematic species** has been used by Grant (1981) to refer to the categories based on fertility relationships as determined by artificial hybridisation experiments. **Ecotype** refers to all members of a species that 'represent a product of genetic response of a species towards a particular habitat'. The ecotypes, which are able to exchange genes freely without loss of fertility or vigour in the offsprings, form an **ecospecies**. An ecospecies corresponds to a taxonomic species. A group of ecospecies capable of limited genetic exchange constitutes a **coenospecies**. A coenospecies is considered equivalent to a subgenus. A group of related coenospecies between which hybridisation is possible directly or through intermediates constitute a **comparium**, which is considered equal to a genus. Complete sterility barriers exist between genera.

Infraspecific Ranks

The species is regarded as the basic unit of classification and many works, including the *Flora of the USSR*, do not recognise infraspecific taxa. Many European, American and Asian Floras, however, do recognise taxa below the rank of species. **The International Code of Botanical Nomenclature** recognises five infraspecific ranks: **subspecies, variety** (Latin, **varieta**), **subvariety, form** (Latin, **forma**) and **subform**. Of these, three (subspecies, variety and form) have been widely used in the literature.

Du Rietz (1930) defined **subspecies** as a *population of several biotypes forming more or less a distinct regional facies of a species*. Morphologically distinct but interfertile populations of a species growing in different geographical regions are maintained as distinct subspecies due to geographical isolation of the species.

Du Rietz defined variety as a *population of several biotypes, forming more or less a local facies of a species*. The term variety is commonly used for morphologically distinct populations occupying a restricted geographical area. Emphasis is on a more localised range of the variety, compared with the large-scale regional basis of a subspecies. Several varieties are often recognised within a subspecies. The term variety is also used for variations whose precise nature is not understood, a treatment often necessary in the pioneer phase of taxonomy.

Forma is often regarded as a sporadic variant distinguished by a single or a few linked characters. Little taxonomic significance is attached to minor and random variations upon which forms are normally based.

Genus

The concept of genus is as old as folk science itself as represented by names such as rose, oak, daffodills, pine and so on. A genus *represents a group of closely related species*. According to Rollins (1953) the function of the genus concept is to bring together species in a phylogenetic manner by placing the closest related species within the general classification. When attempting to place a species within a genus, the primary question would be, is it related to the undoubted species of that genus? Mayr (1957) defined genus as a *taxonomic category which contains either one species or a monophyletic group of species, and is separable from other genera by a decided discontinuity gap*. It was earlier believed that a genus should always be readily definable on the basis of a few technical floral characters. A more rational recognition should take the following criteria into consideration:

1. The group as far as possible should be a natural one. The monophyletic nature of the group should be deduced by cytogenetic and geographic information in relation to morphology.

2. The genera should not be distinguished on a single character but a sum total of several characters. In a number of cases genera are easily recognised on the basis of adaptive characters (adaptations in response to ecological niches) as in the case of establishing aquatic species of *Ranunculus* under a separate genus *Batrachium*.

3. There is no size requirement for a genus. It may include a single species (**monotypic genus**) as *Leitneria* or more than 2000 species as *Senecio*. The only important criterion is that there should be a decided gap between the species of two genera. If the two genera are not readily separable, then they can be merged into one and distinguished as subgenera or sections. Such an exercise should take into consideration the concept of other genera of the family, size of the genus (it is more convenient to have subgenera and sections in a larger genus) and traditional usage.

4. When generic limits are being drawn, it is absolutely necessary that the group of species should be studied throughout the distribution range of the group, because characters stable in one region may break down elsewhere.

Family

A **family** similarly represents a group of closely related genera. Like genus, it is also a very ancient concept because the natural groups now known as families, such as legumes, crucifers, umbels, grasses have been recognised by laymen and taxonomists alike for centuries. Ideally, families should be monophyletic groups. Like the genus, the family may represent a single genus (Podophyllaceae, Hypecoaceae etc.) or several genera (Asteraceae: nearly 1100). Most taxonomists favour broadly conceived family concepts that lend stability to classification. Although there is no marked discontinuity between Lamiaceae (Labiatae) and Verbenaceae, the two are maintained as distinct families. The same tradition prevents taxonomists from splitting Rosaceae, which exhibits considerable internal differences.

6

Phylogeny of Angiosperms

Angiosperms form the most dominant group of plants with at least 234,000 species (Thorne, 1992), a number much greater than all other groups of plants combined together. Not only in numbers, angiosperms are also found in a far greater range of habitats than any other group of land plants. The phylogeny of angiosperms has, however, been a much-debated subject, largely because of very poor records of the earliest angiosperms. These earliest angiosperms probably lived in habitats that were not best suited for fossilisation. Before trying to evaluate the phylogeny it would be useful to have an understanding of the major terms and concepts concerning phylogeny in general, and with respect to angiosperms in particular.

I. IMPORTANT PHYLOGENETIC TERMS AND CONCEPTS

Many important terms have been repeatedly used in discussions on the phylogeny of angiosperms, with diverse interpretation, which has often resulted in different sets of conclusions. A prominent case in point is Melville (1983), who regards the angiosperms as a monophyletic group. His justification, however, that several ancestral forms of the single fossil group Glossopteridae gave rise to angiosperms, renders his view as polyphyletic in the eyes of the greater majority of authors who believe in the strict application of the concept of monophyly. The involvement of more than one ancestor makes angiosperms a polyphyletic group, a view that has been firmly rejected. A uniform thorough evaluation of these concepts is necessary for proper understanding of angiosperm phylogeny.

Plesiomorphic and Apomorphic Characters

A central point to the determination of the phylogenetic position of a particular group is the number of primitive (**plesiomorphic**) or advanced (**apomorphic**) characters that the group contains. In the past most conclusions on primitiveness were based on circular reasoning : 'These families are primitive because they possess primitive characters, and primitive characters are those which are possessed by these primitive families'. Over the recent years a better understanding of these concepts has become possible. It is generally accepted that evolution has proceeded at different rates in different groups of plants so that among the present-day organisms some are more advanced than others. The first step in the determination of relative advancement of characters, is to ascertain which characters are plesiomorphic and which apomorphic. Stebbins (1950) argued that it is wrong to consider the characters as separate entities, since it is through the summation of characters peculiar to an individual that natural selection operates. Sporne (1974) while agreeing with this believed that it is scarcely possible initially to avoid thinking in terms of separate characters, which can be treated better statistically. Given insufficient fossil records of the earliest angiosperms, comparative morphology has been largely used to decide the relative advancement of characters. Many doctrines have been proposed but unfortunately most rely on circular reasoning. Some of the important doctrines are described below:

The **Doctrine of conservative regions** holds that certain regions of plants have been less susceptible to environmental influence than others and therefore exhibit primitive features. Unfortunately, however, over the years every part of the plant has been claimed to be a conservative region. Also, the assumption that a flower is more conservative than vegetative parts is derived from classifications which are based on this assumption.

The **doctrine of recapitulation** holds that early phases in development are supposed to exhibit primitive features, i.e., '*ontogeny repeats phylogeny*'. Gunderson (1939) used it to establish the following evolutionary trends: polypetaly to gamopetaly (since the petal primordia are initially separate; the tubular portion of the corolla arises later), polysepaly to gamosepaly, actinomorphy to zygomorphy and apocarpy to syncarpy. The concept originally applied to animals does not always hold well in plants where ontogeny does not end with embryogeny but continues throughout the adult life. **Neoteny** (persistence of juvenile features in mature organism) is an example wherein a persistent embryonic form represents an advanced condition.

The **doctrine of teratology** was advocated by Sahni (1925). He argued that when normal equilibrium is upset an adjustment is often effected by falling back upon the surer basis of past experience. Thus teratology (abnormality) is seen as reminiscent of some remote ancestor. According to Heslop-Harrison (1952) some teratological phenomena are just likely to be progressive or retrogressive, and each case must be judged on its own merit.

The **doctrine of sequences** advocates that if organisms are arranged in a series in such a way as to show gradation of a particular organ or structure, then the two ends of the series represent apomorphy and plesiomorphy. The most crucial decision, however, is from which end should the series be read.

The **doctrine of association** advocates that if one structure has evolved from another, then the primitive condition of the derived one will be similar to the general condition of the ancestral structure. Thus if vessels have evolved from tracheids then the vessels similar to tracheids (vessels with longer elements, smaller diameter, greater angularity, thinner walls and oblique end walls) represent a more primitive condition than vessels with broader, shorter, more circular elements with horizontal end walls.

The **doctrine of common ground plan** advocates that characters common to all members of a group, must have been possessed by the original ancestor and must therefore be primitive. The doctrine, however, cannot be applied to angiosperms in which there is an exception for almost every character.

The **doctrine of character correlation** was acknowledged during the second decade of the present century when it was realised that certain morphological characters are statistically correlated and the fact can be used in the study of evolution. Sinnot and Bailey (1914) demonstrated a positive correlation between trilacunar node and stipules. Frost (1930) believed that correlation between characters arises because rates of their evolution have been correlated. Sporne (1974) has, however, argued that correlation can be shown to occur even though the rates of evolution of characters are not the same. Within any taxonomic group, primitive characters may be expected to show positive correlation merely because their distribution between the group is not random. By definition, primitive members of that group have retained a relatively high proportion of ancestral (plesiomorphic) characters, while advanced members have dispensed with a relatively high proportion of these same characters either by loss or replacement with different (apomorphic) characters. It follows, therefore, that the distribution of plesiomorphic characters is displaced towards primitive members, which have a higher proportion of plesiomorphic characters, than the average for the group as a whole. Departure from the random can be statistically calculated to establish correlation among characters. Based on these calculations Sporne (1974) prepared a list of 24 characters in Dicotyledons and 14 in Monocotyledons, which show positive correlation. These characters, because of their distribution, have been categorised as **magnoloid** and **amarylloid** respectively. Based on the distribution of these characters, he calculated an **advancement index** for each family and projected the placement of different families of angiosperms in the form of a circular diagram, with most primitive families near the centre, and the most advanced along the periphery. That the earliest members of angiosperms are extinct is clear from the fact that none of the present-day families has the advancement index of zero. All living families have advanced in some respects.

The concept of apomorphic and plesiomorphic characters in understanding the phylogeny of angiosperms has been considerably advanced with recent development of **cladistic methods**. These employ a distinct methodology, somewhat similar to taxometric methods in certain steps involved, leading to construction of **cladograms** depicting evolutionary relationships within a group. Certain groups of angiosperms are reported to have a combination of both plesiomorphic and apomorphic characters, the situation known as **heterobathmy**. *Tetracentron* thus has primitive vesselless wood but the pollen grains are advanced, being tricolpate.

Homology and Analogy

Different organisms resemble one another in certain characters. Taxonomic groups or taxa are constructed based on overall resemblances. The resemblances due to homology are real, whereas those due to analogy are generally superficial. A real understanding of these terms is thus necessary in order to keep organisms with superficial resemblance in separate groups. The two terms as such play a very important role in understanding evolutionary biology.

These terms were first used and defined by Owen (1848). He defined **Homology** as *the occurrence of the same organ in different animals under every variety of forms and functions.* He defined **Analogy** as *the occurrence of a part or an organ in one animal which has the same function as another part or organ in a different animal.* If applied to plants, the rhizome of ginger, the corm of colocasia, tuber of potato, and runner of lawn grass are all homologous, as they all represent a stem. The tuber of potato and the tuber of sweet potato, on the other hand, are analogous as the latter represents a root.

Darwin (1959) was first to apply these terms to both animals and plants. Homology he defined as *that relationship between parts which results from their development from corresponding embryonic parts.* The parts of a flower in different plants are thus homologous, and these in turn are homologous with leaves because their development is identical.

During the latter half of the present century phylogenetic interpretation has been applied to these terms. Simpson (1961) defined homology as *the resemblance due to inheritance from a common ancestry.* Analogy similarly represents *functional similarity and is not due to inheritance from a common ancestry.* Mayr (1969) similarly defined homology as *the occurrence of similar features in two or more organisms, which can be traced to the same feature in the common ancestor of these organisms.* It is thus imperative that homology between two organisms can result only from their having evolved from a common ancestor, and the ancestor must also contain the same feature or features for which the two organisms are homologous.

Wiley (1981) has provided a detailed interpretation of these terms. Homology may either be between two characters, or between two organisms for a particular character. *Two characters are homologous if one is directly derived from the other.* Such a series of characters is called an evolutionary **transformation series** (also called **morphoclines** or **phenoclines**). The original, pre-existing character is termed **plesiomorphic** and the derived one as **apomorphic** or **evolutionary novelty**.

Three or more characters may be homologous if they belong to the same evolutionary transformation series (ovary superior → half-inferior → inferior). The terms plesiomorphic and apomorphic are, however, relative. In an evolutionary transformation series representing characters A, B and C (Fig. 6.1), B is apomorphic in relation to A but it is plesiomorphic in relation to C.

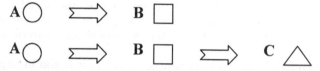

Fig. 6.1 Homology between characters. In the first example character A is plesiomorphic and B is apomorphic. In the second example B is apomorphic in relation to A but plesiomorphic in relation to C as all three belong to an evolutionary transformation series.

Two or more organisms may be homologous for a particular character if their immediate common ancestor also had this character. Such a character is called **shared homologue**.

If the character is present in the immediate common ancestor, but not in the earlier ancestor (Fig. 6.2), i.e. the character is a derived one, the situation is known as **synapomorphy**. If the character is present in the immediate common ancestor, as well as in the earlier ancestor, i.e., it is an original character, the situation is known as **symplesiomorphy** (note sym-). The homology between different organisms is termed **special homology**, as represented by different types of leaves in different species of plants. Different leaves in the same plant such as foliage leaves, bracts, floral leaves would also be homologous, representing **serial homology**. The following criteria may be helpful in identifying homology in practice:

1. Morphological similarity with respect to topographic position, geometric position, or position in relation to other parts. A branch thus occurs in the axil of a leaf, although it may be modified in different ways.
2. Similar ontogeny.
3. Continuation through intermediates as for example evolution of mammalian year from gills of fishes, evolution of achene fruit from follicle in Ranunculaceae. Similarly, vessels having evolved from tracheids, the primitive forms of vessels are more like tracheids, with elongated narrower elements with oblique end walls.

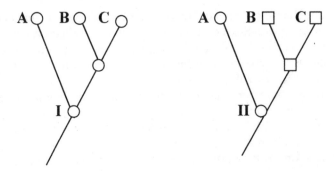

Fig. 6.2 Homology between two organisms B and C. In diagram I similarity is due to symplesiomorphy as the character was unchanged in the previous ancestor. In II it is due to synapomorphy as the previous ancestor had a plesiomorphic character and the two now share a derived character.

4. When the same relatively simple character is found in large number of species, it is probably homologous in all the species. Sets of characters may similarly be homologous.
5. If two organisms share characters of sufficient complexity and judged homologous, other characters shared by the organisms are also likely to be homologous.

Parallelism and Convergence

Unlike homology, if the character shared by two organisms is not traced to a common ancestor the similarity may be the result of **homoplasy**. It can result in two different ways. One, the organisms have a common ancestor but the character was not present in their common ancestor (parallelism). It could also result from two different characters in different ancestors evolving into identical characters (convergence). Both situations represent **false synapomorphy** because the similar character is derived and not traced to a common ancestor.

Simpson (1961) defined **parallelism** as *the independent occurrence of similar changes in groups with a common ancestry, and because they had a common ancestry*. The two species *Ranunculus tripartitus* and *R. hederacea* have a similar aquatic habit and dissected leaves and have acquired these characters by parallel evolution. The development of vessels in Gnetales and dicotyledons also represents a case of parallelism.

Convergence implies *increasing similarity between two distinct phyletic lines, either with regard to individual organ or to the whole organism*. The similar features in convergence arise separately in two or more genetically diverse and not closely related taxa or lineages. The similarities have arisen in spite of lack of affinity and have probably been derived from different systems of genes. Examples may be found in the occurrence of pollinia in

Asclepiadaceae and Orchidaceae, and the 'switch habit' (circular sheath at nodes) in *Equisetum, Ephedra* and *Polygonum.* The concepts of parallelism and convergence are illustrated in Fig. 6.3.

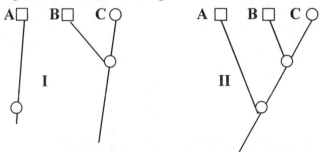

Fig. 6.3 Examples of convergence (I) and parallelism (II) between organisms A and B. In convergence similarity is between organisms derived from different lineages. In parallelism the ancestor is common but both A and B have evolved a plesiomorphic character independently. In both cases similarity represents false synapomorphy. Dissimilarity between B and C in both diagrams is due to divergence.

Convergence is generally brought about by similar climates and habitats, similar methods of pollination or dispersal. Once the convergence has been identified between two taxa, which have been grouped together, they are separated to make the groups natural and monophyletic. The following criteria may help in the identification of convergence:

1. Convergence commonly results from **adaptation to similar habitats**. Water plants thus usually lack root hairs and root cap but contain air lacunae; annuals are predominant in deserts, which also have a good number of succulent plants. The gross similarity between certain succulent species of Euphorbiaceae and Cactaceae is a very striking example of convergence.
2. Convergence may also result from **similar modes of pollination** such as wind pollination in such unrelated families as Poaceae, Salicaceae and Urticaceae, pollinia in Asclepiadaceae and Orchidaceae.
3. Convergence may also be due to **similar modes of dispersal** as seen in hairy seeds of Asteraceae, Asclepiadaceae and some Malvaceae.
4. Convergence commonly occurs between relatively advanced members of respective groups. *Arenaria* and *Minuartia* form natural groups of species which were earlier placed within the same genus *Arenaria*. The two species *Arenaria leptocladus* and *Minuartia hybrida* show more similarity than between any two species of these two genera. If the similarity is **patristic** (result of common ancestry), and the two species represent the most primitive members of their respective groups (Fig. 6.4-I) and it would have been advisable to place all the species in the same genus *Arenaria*. Studies have shown, however, that these two species

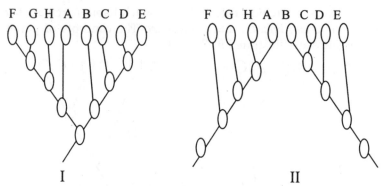

Fig. 6.4 Two possible reasons for similarity between species A and B. In (I) A (cf. *Arenaria leptocladus*) and B (cf. *Minuartia hybrida*) are the most primitive members of respective lineages FGHA (cf. *Arenaria*) and BCDE (cf. *Minuartia*). The two lineages share a common ancestry and thus constitute a single monophyletic group (cf. *Arenaria* s. l.). In (II) A and B happen to be the most advanced members of the respective groups; the two lineages are distinct and as such similarity between A and B is superficial due to convergence, justifying the independent recognition of two lineages (cf. distinct genera *Arenaria* and *Minuartia*).

are the most specialised in each group (Fig. 6.4-II) and thus show convergence. Separation of the two genera is therefore justified, because placing all the species within the same genus *Arenaria* would render the group polyphyletic, a situation that evolutionary biologists avoid.

Monophyly, Paraphyly and Polyphyly

These terms have been commonly used in taxonomy and evolutionary literature with such varied interpretation that much confusion has arisen in their application. Defined broadly, the terms monophyly (derivation from a single ancestor) and polyphyly (derivation from more than one ancestors) would differ in meaning depending upon how far back we are prepared to go in evolutionary history. If life arose only once on Earth, all organisms (even if you place an animal species and a plant species in the same group) are ultimately monophyletic in origin. There is thus a need to precise these terms, to make them meaningful in taxonomy.

Simpson (1961) defined **monophyly** as *the derivation of a taxon through one or more lineages from one immediately ancestral taxon of the same or lower rank*. Such a definition would be true if say genus B evolved from genus A through one species of the latter, since in that case the genus would be monophyletic at the same rank (genus) as well as at the lower (species) rank. On the other hand, if genus B evolved from two species of genus A, it would be monophyletic at the genus level but polyphyletic at the lower rank. Most authors, however, including Heslop-Harrison (1958) and Hennig (1966), adhere to a stricter interpretation of monophyly, namely the group should have evolved from a single immediately ancestral species belonging to the group in question. There are thus two different levels of monophyly: a **minimum monophyly** wherein one supraspecific taxon is derived from another of equal rank (Simpson's definition), and a **strict monophyly** wherein one higher taxon is derived from a single evolutionary species.

Mayr (1969) and Melville (1983) follow the concept of minimum monophyly. Most authors, including Heslop-Harrison (1958), Hennig (1966), Ashlok (1971) and Wiley (1981), reject the idea of minimum monophyly. All supraspecific taxa are composed of individual lineages that evolve independent of each other and cannot be ancestral to one another. Only a species can be an ancestor of a taxon. The supraspecific ancestors and, for that matter, supraspecific taxa are not biologically meaningful entities and are only evolutionary artefacts.

Hennig (1966) defined a monophyletic group as *a group of species descended from a single ('stem') species, which includes all descendants from this species*. Briefly, a monophyletic group comprises all descendants that at one time belonged to a single species. A useful analysis of Hennig's concept of monophyly was made by Ashlok (1971). He distinguished between two types of monophyletic groups—those that are **holophyletic** when *all descendants of the most recent common ancestor are contained in the group* (monophyletic sensu Hennig) and those that are **paraphyletic** and *do not contain all descendants of the most recent common ancestor of the group*. A **polyphyletic** group according to him is one whose most recent ancestor is not cladistically a member of that group. The terms holophyletic and monophyletic are now considered synonymous. Diagrammatic representations of Ashlok's concept of polyphyly, monophyly and paraphyly are presented in Fig. 6.5.

An excellent representation of monophyly, paraphyly and polyphyly is presented by '**cutting rules**' devised by Dahlgren and Rasmussen (1983). The distinction is based on how the group is separated from a representative evolutionary tree (Fig. 6.6). A **monophyletic group** is *separated by a single cut below the group*, i.e., it represents *one complete branch*. A **paraphyletic group** is *separated by one cut below the group and one or more cuts higher up*, i.e., it represents *one piece of a branch*. A **polyphyletic group**, on the other hand, is *separated by more than one cut below the group*, i.e., it represents *more than one piece of a branch*.

It is a widely accepted principle that taxonomic groups (at least above the species level) should be monophyletic. If it is confirmed that the group is polyphyletic, it is advisable to split it into monophyletic groups. Based on this principle, the genus *Paeonia* has been separated from Ranunculaceae into a distinct family Paeoniaceae. If the group represents paraphyly an

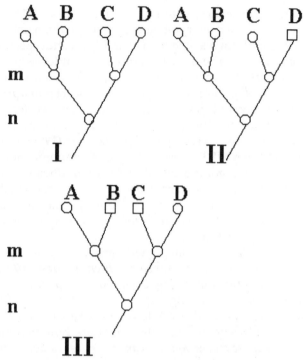

Fig. 6.5 Concepts of monophyly, paraphyly and polyphyly. In (I) groups AB and CD are monophyletic as each has a common ancestor at level m. Similarly group ABCD is monophyletic as it has a common ancestor at level n. In (II) group ABC is paraphyletic as we are leaving out descendant D of the common ancestor at level n. In (III) group BC is polyphyletic as their respective ancestors at level m do not belong to this group.

attempt should be made to identify the remaining members and include these within the group to make it monophyletic. Sosef (1997) compares the existent hierarchical models of classification. He argues that a phylogenetic tree can be subdivided according to a monophyletic hierarchical model, in which only monophyletic units figure or, according to a 'Linnaean' hierarchical model, in which both mono- and paraphyletic units occur. Most present-day phylogeneticists try to fit the monophyletic model within the set of nomenclatural conventions that fit the Linnaean model. However, the two models are intrinsically incongruent. The monophyletic model requires a system of classification of its own, at variance with currently accepted conventions. Since, however, the monophyletic model is unable to cope with reticulate evolutionary relationships, it is unsuited for the classification of nature. The Linnaean model is to be preferred. This renders the acceptance of paraphyletic supraspecific taxa inevitable.

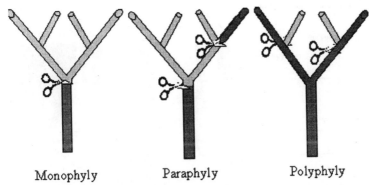

Monophyly   Paraphyly   Polyphyly

Fig. 6.6 Application of cutting rules to distinguish between monophyly, paraphyly and polyphyly. The group is represented by lighter portion of the tree. A monophyletic group can be separated by a single cut below the group, a paraphyletic group by one cut below the group and one or more higher up. A polyphyletic is separated by more than one cut below the group. A monophyletic group represents one complete branch, a paraphyletic group one larger portion of the branch, whereas a polyphyletic group represents more than one piece of a branch.

Phylogenetic tree, Phenogram, Phylogram and Cladogram

The affinities between various groups of plants are commonly depicted with the help of diagrams, with several innovations. An understanding of these terms is necessary for a correct interpretation of putative relationships. These branching diagrams are broadly known as **dendrograms**. The **phylogenetic tree** is a commonly used diagram in relating the phylogenetic history. The vertical axis in such a diagram represents the geological time scale. In such a diagram the origin of a group is depicted by the branch diverging from the main stock and its disappearance by the branch termination. Branches representing the fossil groups end in the geological time when the group became extinct, whereas the living plant groups extend up to the top of the tree. As already mentioned, the relative advancement of the living groups is indicated by their distance from the centre, primitive groups being near the centre, and advanced groups towards the periphery. A phylogenetic tree representing possible relationships and evolutionary history of seed plants is presented in Fig. 6.7.

Dahlgren (1975) presented the phylogenetic tree (preferred to call it phylogenetic 'shrub' in 1977) of flowering plants with all living groups reaching the top, and the cross-section of the top of the phylogenetic tree was shown as top plane of this diagram (Fig. 6.8).

In subsequent schemes by Dahlgren (1977, 1983, 1989), the branching portion of the diagram was dropped and only the top plane (cross-section of the top) presented (Fig. 6.9). Thorne's diagram (1983) is likewise the top

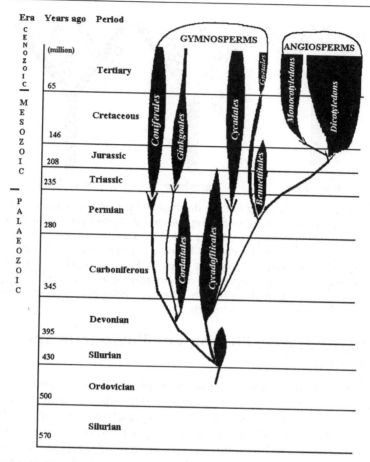

Fig. 6.7 A phylogenetic tree representing the evolutionary history of plants including angiosperms. The vertical axis represents the geological time scale.

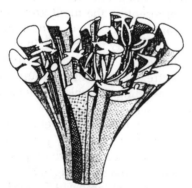

Fig. 6.8 Phylogenetic tree (subsequently named phylogenetic 'shrub' by Dahlgren) of angiosperms presented by Dahlgren with a section of the top (adapted from Dahlgren, 1977).

Fig. 6.9 Dahlgrenogram representing Dahlgren's diagram of a transverse section through the top of a phylogenetic tree for angiosperms as presented by his wife Gertrud Dahlgren (1989) (courtesy Gertrud Dahlgren).

view of a **phylogenetic shrub** (Fig. 6.10) in which the centre representing the extinct primitive angiosperms now absent, is empty.

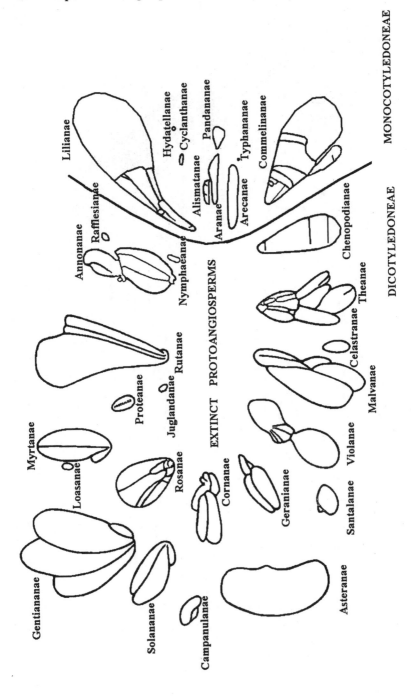

Fig. 6.10   Thorne's phylogenetic shrub of angiosperms (redrawn from Thorne, 1992).

Diagrams in which the vertical axis represents the degree of apomorphy were earlier known as **cladograms** (Stace, 1980), but the term has now been restricted to diagrams constructed through the distinct methodology of cladistics (Stace, 1989). Diagrams with the vertical axis representing the degree of apomorphy are now more appropriately known as **phylograms**. The earliest well-known example of such a phylogram is **'Bessey's cactus'** (see Fig. 2.7). In such diagrams the most primitive groups end near the base and the most advanced reach the farthest distance. Hutchinson (1959, 1973) presented his phylogram in the form of a line diagram. The recent classifications of Takhtajan (1966, 1980, 1987) and Cronquist (1981, 1988) are more innovative phylograms in which the groups are depicted in the form of balloons or bubbles whose size corresponds to the number of species in the group. Such phylograms thus not only depict phylogenetic relationships between the groups, they also show the degree of advancement as also the relative number of species in different groups. Such diagrams have been popularly known as **bubble diagrams**. The bubble diagram of Takhtajan is more detailed and shows the relation of the orders within the 'bubble'; as mentioned earlier, Woodland (1991) aptly described it as **'Takhtajan's flower garden'**.

A **cladogram** represents an evolutionary diagram utilising cladistic methodology, which attempts to find the shortest hypothetical pathway of changes within a group that explains the present phenetic pattern, using the **principle of parsimony**. A cladogram is a representation of the inferred historical connections between the entities as evidenced by synapomorphies. The vertical axis of the cladogram is always an implied, but usually non-absolute time-scale. Cladograms are ancestor-descendant sequences of populations. Each bifurcation of the cladogram represents a past speciation that resulted in two separate lineages.

A **Phenogram** is a diagram constructed on the basis of numerical analysis of phenetic characters. Such a diagram is the result of utilisation of a large number of characters, usually from all available fields, and involves calculating similarity between taxa and constructing a diagram through **cluster analysis**. Such a diagram is very useful, firstly because based on a large number of characters and, secondly because a hierarchical classification can be achieved by deciding upon the threshold levels of similarity between taxa assigned to various ranks.

It must be pointed out, however, that considerable confusion still exists between application of the terms cladogram and phylogenetic tree. Wiley (1981) defines a cladogram as *a branching diagram of entities in which branching is based on inferred historical connections between the entities as evidenced by synapomorphies. It is thus a phylogenetic or historical dendrogram.* He defines a phylogenetic tree as *a branching diagram portraying hypothesised genealogical ties and sequences of historical events linking individual organisms, populations, or taxa.* At the species and population level the number of possible phylogenetic trees could be more than cladograms for particular character changes, depending on which species is ancestral and being relegated lower down on the vertical axis. In the case of higher taxa the number of cladograms and phylogenetic trees could possibly be equal, because higher taxa cannot be ancestral to other higher taxa since they are not units of evolution but historical units composed of separately evolving species.

II. ORIGIN OF ANGIOSPERMS

The origin and early evolution of angiosperms are enigmas that have intrigued botanists for well over a century. They constituted an **'abominable mystery'** to Darwin. The mystery is slowly being 'sleuthed' and at the present pace of Sherlock Holmes' research, may be no more mysterious within the next two decades than for any other major group. With the exception of coniferous forest and moss-lichen tundra, angiosperms dominate all major terrestrial vegetation zones, account for the majority of primary production on land, and exhibit bewildering morphological diversity. Unfortunately, much less is known about the origin and early evolution of angiosperms, resulting in a number of different views regarding their ancestors, the earliest forms and course of evolution. The origin of angiosperms may be conveniently discussed under the following considerations.

What are Angiosperms?

Angiosperms form a distinct group of seed plants sharing a unique combination of characters. The important characters include carpels enclosing the ovules, pollen grains germinating on the stigma, sieve tubes with companion cells, double fertilisation resulting in triploid endosperm, and highly reduced male and female gametophytes. The angiosperms also have vessels. The pollen grains of angiosperms are also unique in having non-laminate endexine and ectexine differentiated into a foot-layer, columellar layer and tectum. The angiosperm flower typically is a hermaphrodite structure with carpels surrounded by stamens and the latter by petals and sepals since insect pollination prevails.

There may be individual exceptions to most of these characters. Vessels are absent in some angiosperms (Winteraceae) and some gymnosperms have vessels (Gnetales). The flowers are unisexual without perianth in several Amentiferae, which also exhibit anemophily.

In spite of these and other exceptions, this combination of characters is unique to angiosperms and found in no other group of seed plants.

What is the Age of Angiosperms?

The time of origin of angiosperms is a matter of considerable debate (Table 6.1). For many years the earliest well-documented angiosperm fossil was considered to be the form-genus *Clavitopollenites* described (Couper, 1958) from Barremian and Aptian strata of Early Cretaceous of southern England (132 to 112 Ma), a monosulcate pollen with distinctly sculptured exine, resembling the pollen of the extant genus *Ascarina*. Brenner and Bickoff (1992) recorded similar but inaperturate pollen grains from the Valanginian (ca 135 Ma) of the Helez formation of Israel, now considered to be the oldest record of angiosperm fossils (Taylor and Hickey, 1996). The number and diversity of angiosperm fossils increased suddenly and by the

Table 6.1: Geological time scale

Time (Ma)	Era	Period	Epoch	Stage
__0.01__		Quaternary	Holocene	
__2.5__			Pleistocene	
__7__	Cenozoic		Pliocene	
__26__			Miocene	
__38__		Tertiary	Oligocene	
__54__			Eocene	
__65__			Palaeocene	
__74__				Maestrichtian
__83__				Campanian
__87__				Santonian
__89__			Upper	Coniacian
__90__				Turonian
__97__				Cenomanian
__112__		Cretaceous		Albian
__125__				Aptian
__132__	Mesozoic			Barremian
__135__			Lower	Hauterivian
__141__				Valanginian
__146__				Berriasian
			Upper	
		Jurassic	Middle	
__208__			Lower	
			Upper	
		Triassic	Middle	
__235__			Lower	
__280__		Permian		
__345__		Carboniferous		
__395__		Devonian		
__430__	Palaeozoic	Silurian		
__500__		Ordovician		
__570__		Cambrian		
__2400__	Precambrian	Algonkian		
__4500__		Archaean		

end of the Early Cretaceous (ca 100 Ma) major groups of angiosperms, including herbaceous Magnoliidae, Magnoliales, Laurales, Winteroids and Liliopsida were well represented. In the Late Cretaceous at least 50% of the species in the fossil flora were angiosperms. By the end of the Cretaceous many extant angiosperm families had appeared. They subsequently increased exponentially and constituted the most dominant land flora, continuing up to the present.

The trail in the reverse direction is incomplete and confusing. Many claims of angiosperm records before the Cretaceous were made but largely rejected. Erdtman (1948) described *Eucommiidites* as a tricolpate dicotyledonous pollen grain from the Jurrassic. This, however, had bilateral symmetry instead of the radial symmetry of angiosperms (Hughes, 1961) and granular exine with gymnospermous laminated endexine (Doyle et al., 1975). This pollen grain was also discovered in the micropyle of seeds of the female cone of uncertain but clearly gymnospermous affinities (Brenner, 1963). Several other fossil pollens from the Jurassic attributed to Nymphaeaceae ultimately turned out to be gymnosperms.

Several vegetative structures from the Triassic were also attributed to angiosperms. Brown (1956) described *Sanmiguilea* leaves from the Late Triassic of Colorado and suggested affinity with Palmae. A better understanding of the plant was given by Cornet (1986, 1989) who regarded it as a presumed primitive angiosperm with features of monocots and dicots. Although its angiosperm venation was refuted by Hickey and Doyle (1977), Cornet (1989) established its angiosperm venation and associated reproductive structures. Our knowledge of this controversial taxon, however, is far from clear.

Marcouia leaves (earlier described as *Ctenis neuropteroides* by Daugherty, 1941) are recorded from the Upper Triassic of Arizona and New Mexico. Its angiosperm affinities are not clear.

Harris (1932) described *Furcula* of the Upper Triassic of Greenland as bifurcate leaves with dichotomous venation. Although it seems to approach dicots in venation and cuticular structure, it has several non-angiospermous characters including bifurcating midrib and blade, higher vein orders with relatively acute angles of origin (Hickey and Doyle, 1977).

Cornet (1993) has described *Pannaulika*, a dicot-like leaf form from the Late Triassic from the Virginia-North Carolina border. It was considered a three-lobed palmately veined leaf. The associated reproductive structures were attributed to angiosperms but it is not certain that any of the reproductive structures were produced by the plant that bore *Pannaulika*. Taylor and Hickey (1996), however, do not accept its angiosperm affinities, largely on the basis of venation pattern, which resembles more that of ferns. Much more information is needed before the Triassic record of angiosperms is established.

Given the inconclusive pre-Cretaceous record of angiosperms, it is largely believed that angiosperms arose in the Late Jurassic or very Early Cretaceous (Taylor, 1981) nearly 130 to 135 Ma ago (Jones and Luchsinger, 1986).

Melville (1983) who strongly advocated his gonophyll theory, believed that angiosperms arose nearly 240 Ma ago in the Permian and took nearly 140 Ma before they spread widely in Cretaceous. The Glossopteridae which gave rise to angiosperms met a disaster in the Triassic and disappeared, this disaster making progress of angiosperms slow until the Cretaceous when their curve entered an exponential phase. This idea has found little favour, however.

There has been increasing realisation in recent years (Troitsky et al., 1991; Doyle and Donoghue, 1993; Crane et al., 1995) to distinguish two dates one in the Triassic when the **stem angiosperms** ('angiophytes' sensu Doyle and Donoghue, 1993 or 'proangiosperms' sensu Troitsky et al., 1991) separated from sister groups (Gnetales, Bennettitales and Pentoxylales) and the second in the Late Jurassic when the crown group of angiosperms (crown angiophytes) split into extant subgroups (Fig. 6.11).

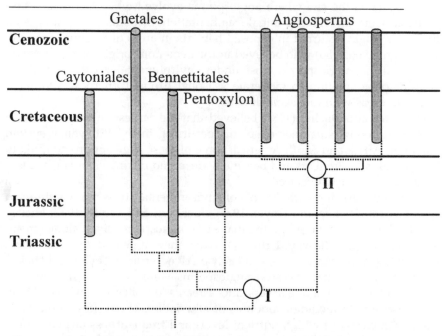

Fig. 6.11 Phylogenetic tree of anthophytes (angiosperm lineage and sister groups). Point (I) marks when angiosperm lineage separated from sister groups in the Late Triassic, and (II) marks the splitting of crown angiosperms into extant subgroups in the Late Jurassic. Dotted line represents conclusions for which fossil record is not available (diagram based on Doyle and Donoghue, 1993).

There have been a number of attempts to estimate divergence times by applying a molecular clock to nucleotide sequence data. The results mostly pointing to much earlier origin of angiosperms have, however, been contradictory. Wolfe et. al. (1989) suggested Late Triassic (200 Ma) as the likely estimate of monocot-dicot split, while Martin et al. (1993) supported Carboniferous origin (~300 Ma) of angiosperms. Sytsma and Baum (1996) conclude that the results strongly caution using the molecular clock for dating until extensive sampling of taxa and genes with quite different molecular evolution is completed. Thus the resolution of angiosperm phylogeny may have to wait for more complete molecular data and its proper appraisal.

What is the Place of Their Origin?

It was earlier believed that angiosperms arose in the Arctic region (Seward, 1931), with subsequent southwards migration. Axelrod (1970) suggested that flowering plants evolved in mild uplands (**upland theory**) at low latitudes. Smith (1970) located the general area of South-East Asia, adjacent to Malaysia as the site where angiosperms evolved when Gondwana and Laurasia were undergoing initial fragmentation. Stebbins (1974) suggested that their origin occurred in exposed habitats in areas of seasonal drought. Takhtajan (1966,1980) who believed in the neotenous origin of angiosperms suggested that angiosperms arose under environmental stress, probably as a result of adaptation to moderate seasonal drought on rocky mountain slopes in areas with monsoon climate.

Retallack and Dilcher (1981) believed that the earliest angiosperms were probably woody, small-leaved plants occurring in the Rift valley system adjoining Africa and South America. Some of these angiosperms adapted to the coastal environments and became widespread during changing sea levels during the Early Cretaceous.

Although agreeing with the role of environmental stress many authors in recent years (Hickey and Doyle, 1977; Upchurch and Wolfe, 1987; Hickey and Taylor, 1992) have suggested that early angiosperms lived along stream and lake margins (**lowland theory**). Later they appeared in more stable backswamp and channel sites, and lastly on river terraces. Taylor and Hickey (1996) suggested that ancestral angiosperms were perennial rhizomatous herbs and evolved along rivers and streams on sites of relatively high disturbance with moderate amounts of alluviation. These sites would have been characterised by high nutrient levels and frequent loss of plant cover due to periodic disturbances.

Angiosperms are Monophyletic or Polyphyletic?

Engler (1892) considered angiosperms to be polyphyletic, monocotyledons and dicotyledons having evolved separately. Considerable diversity of angiosperms in the Early Cretaceous and the extant angiosperms led several authors including Meeuse (1963) and Krassilov (1977) to develop models for polyphyletic origin of angiosperms. This view is largely supported by considerable diversity in the early angiosperm fossils.

Most recent authors, including Hutchinson (1959, 1973), Cronquist (1981, 1988), Thorne (1983, 1992), Dahlgren (1980, 1989) and Takhtajan (1987, 1997) believe in the monophyletic origin of angiosperms, monocotyledons having evolved from primitive dicotyledons. This view is supported by a unique combination of characters such as closed carpels, sieve tubes, companion cells, four microsporangia, triploid endosperm, 8-nucleate embryo sac and reduced gametophytes. Sporne (1974) on the basis of statistical studies also concluded that it is highly improbable that such a unique combination of characters could have arisen more than once independently from gymnosperm ancestors.

It is interesting to note that Melville (1983) considered angiosperms to be monophyletic but the explanation that he offers clubs him with the proponents of polyphyletic origin. He believes that angiosperms arose from several different genera of Glossopteridae. According to him, the species is not always to be considered the ancestor for determining a monophyletic nature. A species from a species is monophyletic, as is a genus from a genus, a family from a family. The principle, according to him, is that to be monophyletic a taxon of any rank must be derived solely from another taxon of the same rank. Glossopteridae and Angiospermidae belong to the same rank subclass. Both taxa consist of minor lineages that may be likened to a rope with many strands, a situation called **pachyphyletic**. This explanation, however, conforms to the concept of minimum monophyly and does not satisfy the rule of strict monophyly, which is now the accepted criterion for monophyly.

What are the Possible Ancestors?

Ancestry of angiosperms is perhaps one of the most controversial and vigorously debated topics. In the absence of direct fossil evidence, almost all groups of fossil and living gymnosperms have been considered as possible ancestors by one authority or the other. Some authors even suggested the **Isoetes origin** of monocotyledons because the plant has a superficial resemblance with onion, albeit no trace of seed habit. The various theories have revolved around two basic theories, viz. the Euanthial theory and the Pseudanthial theory of angiosperm origin. Some other theories have recently received attention also:

1. **Euanthial theory (Anthostrobilus theory).** This theory was first proposed by Arber and Parkins (1907) in which the angiosperm flower is interpreted as being derived from an unbranched bisexual strobilus bearing spirally arranged ovulate and pollen organs, similar to the hermaphrodite

reproductive structures of some extinct bennettitalean gymnosperms. The carpel is thus regarded as a modified megasporophyll (**phyllosporous** origin of carpel). The bisexual strobilus of Magnoliales has been considered to have evolved from such a structure. Also agreeing with this general principle, various authors have tried to identify different gymnosperm groups as possible angiosperm ancestors:

(i) **Bennettitales:** The group appeared in the Triassic and disappeared in the Cretaceous. Lemesle (1946) considered the group to be ancestral to angiosperms, primarily because of the hermaphrodite nature of *Cycadeoidea* which had an elongated receptacle with perianth-like bracts, a whorl of pollen-bearing microsporophylls surrounding the ovuliferous region having numerous ovules and interseminal scales packed together. There were, however, signs of abscission at the base of the male structure which would have shed, exposing the ovular region. The ovules were terminal in contrast to their position in carpels of angiosperms.

(ii) **Caytoniaceae:** Opinion has strongly inclined in recent years towards the probability that angiosperms arose from **Pteridosperms** or **seed ferns**, often placed in the order Lyginopteridales. Caytoniaceae was described from the Jurassic of Yorkshire and Greenland by Harris (1932). The group appeared in the Late Triassic and disappeared towards the end of the Cretaceous. The leaves were borne on twigs and not the trunk. The leaves had two pairs of leaflets and were net veined. Male structures had rachis with branching pinnae, each with a synangium of four sporangia. The seed-bearing structure had rachis with two rows of reflexed cupules. The discovery of pollen grains within the ovules was thought to suggest their true gymnosperm position, however, rather than angiosperm ancestors. Krassilov (1977) and Doyle (1978) regarded the cupule as homologous to the carpel whereas Gaussen (1946) and Stebbins (1974) considered it the outer integument of the ovule. Cladistic studies of Doyle and Donoghue (1987) support caytoniales-angiosperm lineage. Thorne (1996) agreed that angiosperms probably evolved during the Late Jurassic from some group of seed ferns.

(iii) **Cycadales:** Sporne (1971) suggested possible links between Cycadales and angiosperms in the palm-like habit of Cycadales, ovules being borne on leaf-like microsporophylls, trends in the reduction of sporophyll blade being seen in various species of *Cycas*. Although it may be difficult to assume Cycadales as ancestral to angiosperms, the fact that they have been derived from Pteridosperms, and yet resemble angiosperms, further supports the origin of angiosperms from Pteridosperms.

2. **Pseudanthial theory.** Commonly associated with the Englerian School, the theory was first proposed by Wettstein (1907) who postulated that

angiosperms were derived from the gymnosperm order Gnetales, represented by *Ephedra, Gnetum* and *Welwitschia*. The group shows more angiosperm characteristics than any other group of living or fossil gymnosperms. These include presence of vessels, reticulate dicot-like leaves (*Gnetum*), male flower with perianth and bracts, strong gametophyte reduction, and fusion of the second male gametophyte with the ventral canal nucleus. *Ephedra* resembles *Casuarina* in habit. Wettstein homologised the compound strobili of Gnetales with the inflorescences of wind-pollinated Amentiferae, and regarded the showy insect pollinated bisexual flowers of *Magnolia* as pseudanthia derived by aggregation of unisexual units, the carpel thus representing a modified branch (**Stachyosporous** origin of carpel). A number of features, however, refute this theory: different origin of vessels (Bailey, 1944) in angiosperms (from tracheids with scalariform pitting) and Gnetales (from tracheids with circular pitting), several vesselless living angiosperms (cf. Winteraceae) and Amentiferae are now regarded as advanced due to floral reduction. Tricolpate pollen grains also represent an advanced condition. More importantly, Gnetales is a very young group.

But this theory has been strongly supported by Young (1981) who challenged the view that angiosperms were originally vesselless and assumed that vessels were lost in several early lines. Muhammad and Sattler (1982) found scalariform perforations in vessel elements of *Gnetum*, suggesting that angiosperms may be derived from Gnetales after all. Carlquist (1996), however, concludes that this claim from *Gnetum* does not hold when large samples are examined.

The basal group of angiosperms according to this theory included amentiferous-hamamelid orders Casuarinales, Fagales, Myricales and Juglandales. It is significant to note that Wettstein (1907) also included in this basal group, Chloranthaceae and Piperaceae, which have been inviting considerable attention in recent years.

The importance of Gnetales in angiosperm phylogeny has been further strengthened by the discovery of *Welwitschia*-like fossil described by Cornet (1996) as *Archaestrobilus cupulanthus* from the Late Triassic of Texas. The plant had similarly constructed male and female spikes, each possessing hundreds of spirally arranged macrocupules. Male spikes were borne in clusters of three, whereas female spikes occurred singly. Each female macrocupule contained an ovule surrounded by sterile scales. Three to four very small bracts were present and attached near the base of the macrocupule. Each male macrocupule contained sterile scales within. Outside the macrocupule was crowded with numerous bivalved microsporophylls, each with four pollen sacs attached to an inflated stalk. On the outside of the female macrocupule were gland-like structures resembling the stalks bearing pollen sacs on the male macrocupule. Similarly there were sterile filamentous appendages inside the male macrocupule that resemble the sterile scales around the ovule. This suggests an origin from a bisexual macrocupule. The pollen grains are radially symmetrical and monosulcate. The plant is regarded as a gnetophyte more primitive than extant Gnetales.

3. **Anthocorm theory**. This theory is a modified version of the pseudanthial theory and was proposed by Neumayer (1924) and strongly advocated by Meeuse (1963, 1972). According to this theory the angiosperm flower ('functional reproductive unit') has several separate origins (i.e., angiosperms are polyphyletic). In most Magnoliidae and their dicotyledonous derivatives they are modified pluriaxial systems (**holanthocorms**) that originated from the gnetopsids via the Piperales, whereas the modification of an originally uniaxial system (**gonoclad** or **anthoid**) gave rise to flowers of Chloranthaceae. Meeuse (1963) postulated a separate origin of monocotyledons from the fossil order Pentoxylales through the monocot order Pandanales.

Pentoxylales were described from the Jurassic of India and New Zealand. The stem (*Pentoxylon*) had five conducting strands. The pollen-bearing organ (*Sahnia*) was pinnate free above and fused into a cup below. The seed-bearing structure was similar to a mulberry with about 20 sessile seeds, each having an outer fleshy sarcotesta and the inner hard sclerotesta. The sarcotesta was considered homologous to the cupule of seed ferns. The carpel of angiosperms was regarded as a composite structure being an ovule-bearing branch fused with a supporting bract. It is interesting to note, however, that Taylor and Hickey (1996) no longer include *Pentoxylon* as a member of **Anthophytes**, which include angiosperm lineage and its sister groups bennettitales and gnetopsids (see Fig. 6.12). According to them *Pentoxylon* lacks key anthophyte characteristics such as distal, medial and proximal positioning of female, male and sterile organs on the reproductive axis, as well as the enclosure of ovules by bract-derived organs.

4. **Gonophyll theory**. The theory was developed by Melville (1962, 1963, 1983) largely on the basis of a study of a venation pattern. He derived angiosperms from Glossopteridae, which formed important elements in the flora of Gondwanaland. He derived angiosperm from **gonophyll**—a fertile branching axis adnate to a leaf. In simple Glossopterids *Scutum* and *Ottokaria* the fertile branch consisted of a bivaled scale (having two wings) called the scutella with terminal ovules on dichotomous groups of branches. Folding of the scutella along the cluster of its ovules forms the angiosperm condition, an indication of this closure being found in the Permian fossil *Breytenia*. In *Glossopteris* the fertile region is cone-like with a transition from leaves to fertile scales, spirally arranged and conforming to the **anthostrobilus**. In *Mudgea* there is a suggestion of **anthofasciculi**, i.e., leafy structures with two fertile branches, one male, the other female, forming the angiosperm flowers as found in *Ranunculus* and *Acacia*.

Melville believed that angiosperms arose 240 Ma ago in the Permian and took around 140 Ma before they spread widely in the Cretaceous. It is pertinent to point out, as explained earlier that although he considered angiosperms to be monophyletic, his justification puts him among the proponents of the polyphyletic origin of angiosperms.

5. **Herbaceous origin hypothesis**. This hypothesis resembles the Pseudanthial theory but the ancestral plant is considered to be a perennial rhizomatous herb instead of a tree. The term **palaeoherb** was first used by Donoghue and Doyle (1989) for a group of derivative (not ancestral) forms of Magnoliidae having anomocytic stomata, two whorls of perianth and trimerous flowers, including Lactoridaceae, Aristolochiaceae, Cabombaceae, Piperales, Nymphaeaceae and monocots. Taylor and Hickey (1992), while developing their **palaeoherb hypothesis** used the term palaeoherb for the ancestral angiosperm group to exclude monocots and non-herbaceous Lactoridaceae and instead include Chloranthaceae. This ancestral group was given the name 'eoangiosperms' by Hickey and Taylor (1996) to distinguish it from the original concept of palaeoherbs, and including families Chloranthaceae, Saururaceae, Piperaceae, Aristolochiaceae, Barclayaceae, Cabombaceae, Nymphaeaceae and Ceratophyllaceae. They also renamed their model (Taylor and Hickey, 1996) the **Herbaceous Origin hypothesis** and provided ample evidence in support of their conclusions.

According to this hypothesis ancestral angiosperms were small herbaceous plants with a rhizomatous to scrambling perennial habit. They had simple leaves that were reticulately veined and had a primary venation pattern that would have been indifferently pinnate to palmate, whereas the secondary veins branched dichotomously. The vegetative anatomy included sieve-tube elements and elongate tracheary elements with both circular-bordered and scalariform pitting and oblique end walls. The flowers occurred in cymose to racemose inflorescences. The small monosulcate pollen had perforate to reticulate sculpturing. Carpels were free, ascidiate (ovules attached proximally to the closure) with one or two orthotropous, bitegmic, crassinucellate ovule and dicotyledonous embryo. The aforesaid authors cite extreme rarity of fossil angiosperm wood and abundance of leaf impressions in early fossils.

Consensus is emerging from recent phylogenetic studies that gnetopsids represent the closest living relatives of angiosperms, whereas the closest fossil group is the Bennettitales. Angiosperm lineage together with these two groups constitutes **Anthophytes**. The group is believed to have split in the Late Triassic, the angiosperm lineage continuing as **Angiophytes** up to the Late Jurassic when it split into **stem Angiophytes** (the early extinct angiosperms) and **crown Angiophytes** constituting the extant groups of angiosperms (Fig. 6.12).

Krassilov who believed in the polyphyletic origin of angiosperms identified three Jurassic groups as proangiosperms: Caytoniales, Zcekanowskiales and Dirhopalostachyaceae. Pollen germinating on the lip, according to him would be rather disappointing because these plants would then be classified as angiosperms and excluded from discussion of their ancestors. He evolved the Laurales-Rosales series from Caytoniales. Zcekanowskiales had bivalved capsules provided with stigmatic bands and showed links with

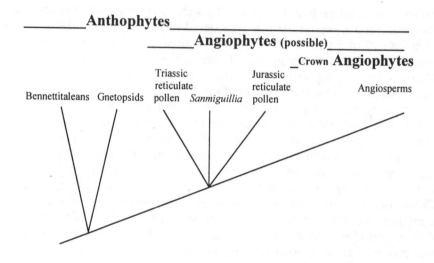

Fig. 6.12 A consensus phylogeny of Anthophytes proposed by Taylor and Hickey (1996). Note that *Pentoxylon* has been excluded from sister groups (now only Bennettitales and Gnetopsids) of Angiophytes.

monocots. Dirhopalostachyaceae had paired ovules exposed on shield-like lateral appendages and probably evolved in Hamamelidales.

Origin of Monocotyledons

It was originally believed (Engler, 1892) that monocotyledons arose before dicotyledons and are polyphyletic (Meeuse, 1963). It is now largely believed that monocotyledons evolved from dicots monophyletically. According to Bailey (1944) and Cheadle (1953) vessels had independent origin and specialisation in monocots and dicots, and thus monocots arose from vesselless dicots. Cronquist did not agree with the independent origin of vessels in two groups. He considered monocots to have an aquatic origin from ancestors resembling modern Nymphaeales. This was strongly refuted, however, by studies of vessels done by Kosakai, Mosely and Cheadle (1970). They considered it difficult to believe that putatively primitive Alismataceae evolved advanced vessels in an aquatic environment yet gave rise to terrestrial monocots with more primitive vessel elements in the metaxylem of roots. They thus favoured the origin of Alismataceae from terrestrial forms.

According to Hutchinson (1973), monocots arose from Ranales along two lines, one (Ranunculoideae) giving rise to Alismatales and other (Helleboroideae) giving rise to Butomales. Takhtajan (1980, 1987) proposed a common origin for Nymphaeales and Alismatales from a hypothetical terrestrial herbaceous group of Magnoliidae. Dahlgren et al. (1985) believed that monocots appeared in the Early Cretaceous some 110 Ma ago when the ancestors of Magnoliiflorae must have already acquired some of the present attributes of that group but were less differentiated; some other dicotyledonous groups had already branched off from the ancestral stock. Thorne (1996) believes that monocotyledons appear to be very early off-shoot of the most primitive dicotyledons.

Most Primitive Living Angiosperms

There was general agreement for nearly a century that the early angiosperms were woody shrubs or small trees (herbaceous habit being derived), with simple evergreen entire and pinnately veined leaves with stipules. Concerning the most primitive living angiosperms there have been two opposing points of view.

According to the **Englerian School,** a view now largely rejected, Amentiferae (including Juglandaceae, Betulaceae, Fagaceae etc.) with reduced unisexual flowers in catkins (or aments) constitute the most primitive living angiosperms. It is now agreed that Amentiferae have advanced tricolpate pollen grains, wood anatomy is relatively advanced and the simplicity of flowers is due to reduction rather than primitiveness. They have also secondarily achieved wind pollination.

The alternative **Besseyan School (Ranalian School)** considers the Ranalian complex (including Magnoliales), having bisexual flowers with free, equal, spirally arranged floral parts, representative of most primitive angiosperms. Bessey (1915), Hutchinson (1959, 1973), Takhtajan (1966, 1969) and Cronquist (1968) have all proposed that the large solitary flower of *Magnolia* (Fig. 6.13) ('Magnolia the primitive theory') with an elongated floral axis

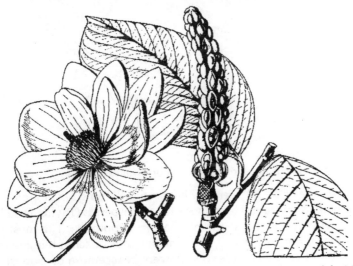

Fig. 6.13 Flower and a twig of *Magnolia campbellii* with elongated fruiting axis.

bearing numerous spirally arranged stamens and carpels, is the most primitive living representative. However, this view was later, challenged by Gottsberger (1974) and Thorne (1976), who considered the most primitive flowers to have been moderate in size, with fewer stamens and carpels, and grouped in lateral clusters as in the family Winteraceae, to which such primitive genera as *Drimys* (Fig. 6.14) have been assigned. This view is supported by the occurrence of similar stamens and carpels, absence of vessels, morphology similar to Pteridosperms, and high chromosome number suggesting a long evolutionary history, and less specialised beetle pollination of *Drimys* compared to *Magnolia*. Takhtajan (1980, 1987) later acknowledged that the moderate sized flowers of *Degeneria* and Winteraceae are primitive and the large flowers of *Magnolia* and Nymphaeaceae are of secondary origin. However, he considered Degeneriaceae to be the most primitive family of living angiosperms. Cronquist (1981, 1988) also discarded *Magnolia* but considered Winteraceae to be the most primitive. Suggestions have also come projecting Calycanthaceae (Loconte and Stevenson, 1991) as basic angiosperms with a series of vegetative and reproductive angiosperm plesiomorphies such as shrub habit, unilacunar two-trace nodes, opposite leaves, strobilar flowers, leaf-like bracteopetals and few ovulate carpels.

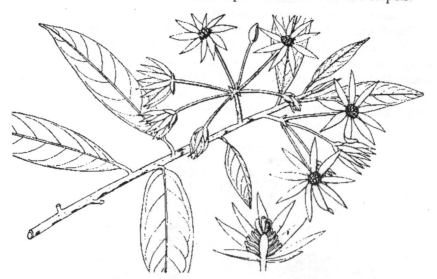

Fig. 6.14 Flowering twig of *Drimys winteri*.

The last decade of the 20th century has seen the strong development of an alternative herbaceous origin hypothesis for angiosperms (Taylor and Hickey, 1996), originally developed as the palaeoherb hypothesis. The most primitive angiosperms are considered to be rhizomatous or scrambling perennial herbs with simple net-veined leaves, flowers in racemose or cymose inflorescences, with free carpels containing one or two ovules. Taylor and Hickey

(1996) consider Chloranthaceae the basic angiosperm family. Chase et al. (1993) had earlier expressed the view that Ceratophyllaceae represents the basic angiosperm family. Cladistic studies by Sytsma and Baum (1996) based on molecular data support placement of *Ceratophyllum* at the base of angiosperms, but the authors cautioned that resolution of basal angiosperm relationships may have to await both the collection of additional molecular and morphological data as well as further theoretical advances in phylogenetic systematics.

Evolution within Angiosperms

Although there has been some recent controversy regarding the habit of the most primitive living angiosperms being woody or herbaceous, the general features of primitive angiosperms are largely settled. They have simple alternate exstipulate leaves, which are entire and petiolate with poorly organised reticulate venation and with unilacunar, two-trace nodes. The vessels are absent or tracheid-like. Flowers are bisexual, radially symmetrical with spirally arranged floral parts. Stamens are broad, undifferentiated with marginal microsporangia. Carpels are broad with large number of ovules, stigma along the margin and not completely sealed, ovules bitegmic, crassinucellate. Fruits are follicular. From such a primitive conditions have developed several phyletic lines showing various evolutionary trends as underlined below.

1. **Fusion.** During the course of evolution in angiosperms fusion of different parts has led to floral complexity. Fusion of like parts has led to the development of gamosepaly, gamopetaly, synandry and syncarpy in various families of angiosperms. Stamens have shown fusion to different degrees: fusion of filaments only (monadelphous condition in Malvaceae), fusion of anthers only (syngenesious condition found in Asteraceae) or complete fusion (synandry as in *Cucurbita*). Carpels may similarly be fused only by ovaries (Caryophyllaceae), only by styles (Apocynaceae) or complete fusion (Solanaceae, Primulaceae). Fusion of unlike parts has resulted in an epipetalous condition (fusion of petals and stamens) and formation of an inferior ovary (fusion of calyx with ovary).

2. **Reduction.** Relatively simple flowers of many families have primarily been the result of reduction. The loss of either stamens or carpels has resulted in unisexual flowers. The loss of one perianth whorl has resulted in monochlamydeous forms, and their total absence in achlamydeous forms. There has also been individual reduction in the number of perianth parts, number of stamens and carpels. Within the ovary different genera have shown reduction in the number of ovules to ultimately one, as seen in the transformation of follicle into achene within the family Ranunculaceae. There has also been reduction in size of flowers, manifested in diverse families such as Asteraceae and Poaceae. Reduction in size of seeds has been extreme in Orchidaceae.

3. **Change in symmetry**. From simple radially symmetrical actinomorphic flowers in primitive flowers developed zygomorphic flowers in various families to suit insect pollination. The size of corolla tube and orientation of corolla lobes changed according to the mouthparts of the pollinating insects, with striking specialisation achieved in the turn pipe mechanism of *Salvia* flowers, and female wasp-like flowers of orchid *Ophrys*.

4. **Elaboration**. This compensating mechanism has been found in several families. In Asteraceae and Poaceae, the reduction in size of flowers has been compensated by increase in number of flowers in the inflorescence. Similarly, reduction in number of ovules has been accompanied by increase in size of ovule and ultimately seed as seen in *Juglans* and *Aesculus*.

5. **Remoration**. The term was suggested by Melville (1983) to refer to evolutionary retrogression found in angiosperms and their fossil relatives. The fertile shoots of angiosperms, according to him, show venation pattern changes progressively from vegetative leaves through successive older evolutionary stages in bracts and sepals, and the most ancient in petals. The innermost parts in a bud as such represent the most primitive evolutionary condition, and the outermost the most recent condition.

There has been some shift in the understanding of angiosperm phylogeny to support stachyosporous origin of angiosperm carpel (Taylor and Kirchner, 1996). With the acceptance of such a viewpoint, the reproductive axis with many flowers, few carpels per flower and few ovules per carpel are ancestral. Evolution proceeded along two directions from this: one with few flowers, each of which had many carpels and few ovules and the other with few flowers, each containing few carpels and many ovules. The evolutionary trends in angiosperms are thus often complicated and frequent reversal of trends may be encountered, as for example the secondary loss of vessels in some members.

Carlquist (1996), based on a survey of wood anatomy, identified a number of distict evolutionary trends in angiosperms. The cambial intitials have shortened, the ratio of length accompanying fibers to vessel elements (F/V ratio) has shown an increase from 1.00 in primitive dicotyledons to about 4.00 in the most specialised woods, and angular outline of vessels changed to circular outline together with widening of their diameter. It is interesting to note that cladistic studies have shown that present-day vesselless angiosperms do not form a single clade and are distributed in diverse groups such as Hamamelidales (*Trochodendron* and *Tetracentron*), Magnoliales (Amoborellaceae, Winteraceae), and Laurales (*Sarcandra*: Vessels have been reported, however, in root secondary xylem by Carlquist, 1987 and in stem metaxylem by Takahashi, 1988). This led Carlquist to conclude that vessels have originated numerous times in dicotyledons.

7

Major Systems of Classification

BENTHAM AND HOOKER

The system of classification of seed plants presented by Bentham and Hooker, two English botanists, represented the most well-developed natural system. The classification was published in a three-volume work *Genera plantarum* (1862-83). George Bentham (1800-1884) was a self-trained botanist (Fig. 7.1).

Fig. 7.1 George Bentham (1800-1884), coauthor of *Genera plantarum* (with J.D. Hooker, 1862-1883) and the author of the 7-volume *Flora Australiensis* and several monographs on major families (reproduced with permission from Royal Botanic Gardens, Kew).

He was extremely accomplished and wrote many important monographs on families such as Labiatae, Ericaceae, Scrophulariaceae and Polygonaceae. He also published *Handbook of British Flora* (1858) and *Flora Australiensis* in 7 volumes (1863-78). Sir J.D. Hooker (1817-1911), who succeeded his father William Hooker as Director, Royal Botanic Gardens in Kew, England was a very well-known botanist, having explored many parts of the world (Fig. 7.2). He published *Flora of British India* in 7 volumes (1872-97), *Student's Flora of the British Isles* (1870) and also revised later editions of *Handbook of British Flora*, which remained a major British Flora until 1952. He also supervised the publication of *Index Kewensis* (2 volumes, 1893), listing names of all known species and their synonyms.

Fig. 7.2 Sir Joseph Dalton Hooker (1817-1911), the famous British botanist who co-authored *Genera plantarum* with George Bentham, besides authoring the 7-volume *Flora of British India* and several other publications. He was the Director of the Royal Botanic Gardens, Kew (reproduced with permission from Royal Botanic Gardens, Kew).

The *Genera plantarum* of Bentham and Hooker provided the classification of seed plants, describing 202 families and 7569 genera. They estimated the seed plants to include 97,205 species. The classification was a refinement of the systems proposed by A.P. de Candolle and Lindley, which in turn were based on that of de Jussieu. The delimitation of families and genera was based on natural affinities and was pre-Darwinian in concept. The descriptions were based on personal studies from specimens and not a mere compilation of known facts, an asset which made the classification so popular and authentic. Many important herbaria of the world have specimens arranged according to this system.

The system divided Phanerogams or seed plants into three classes: **Dicotyledons, Gymnospermae** and **Monocotyledons.**

Dicotyledons were further subdivided into three subclasses: **Polypetalae, Gamopetalae** and **Monochlamydeae** based on the presence or absence of petals and their fusion. These were further subdivided into series, orders (called cohorts by the two authors) and families (called natural orders). No orders (cohorts) were recognised within Monochlamydeae and Monocotyledons, the series being directly divided into families (natural orders). A broad outline of the classification is presented in Table 7.1

Table 7.1: Outline of the system of classification presented by Bentham and Hooker in *Genera plantarum* (1862-1883)

Phanerogams or seed plants
Class 1. Dicotyledons—*14 series, 25 orders, 165 families*
Subclass 1. Polypetalae
              Series 1. Thalamiflorae...*6 orders*
                       2. Disciflorae...*4 orders*
                       3. Calyciflorae...*5 orders*
  Subclass 2. Gamopetalae
              Series 1. Inferae...*3 orders*
                       2. Heteromerae...*3 orders*
                       3. Bicarpellatae...*4 orders*
  Subclass 3. Monochlamydeae
              Series 1. Curvembryeae
                       2. Multiovulatae aquaticae
                       3. Multiovulatae terrestres
                       4. Microembryeae
                       5. Daphanales
                       6. Achlamydosporae
                       7. Unisexuales
                       8. Ordines anomali
Class 2. Gymnospermae—*3 families*
Class 3. Monocotyledons—*7 series, 34 families*
              Series 1. Microspermae
                     2. Epigynae
                     3. Coronarieae
                     4. Calycinae
                     5. Nudiflorae
                     6. Apocarpae
                     7. Glumaceae

Merits

The fact notwithstanding that the system does not incorporate phylogeny and is more than 100 years old, it still enjoys a reputation of being a very sound system of classification, owing to the following merits:

1. The system has great practical value for identification of plants. It is very easy to follow for routine identification.
2. The system is widely followed for the arrangement of specimens in the herbaria of many countries including Britain and India.
3. The system is based on a careful comparative examination of actual specimens of all living genera of seed plants and is not a mere compilation of known facts.
4. Unlike de Candolle, the **Gymnosperms** are not placed among dicots but rather in an independent group.
5. Although the system is not a phylogenetic one, **Ranales** are placed in the beginning of Dicotyledons. The group Ranales (in the broader sense including families now separated under order Magnoliales) is generally regarded as **primitive** by most of the leading authors.
6. Dicotyledons are placed before the Monocotyledons, a position approved by all present-day authors.
7. The description of families and genera are precise. Keys to the identification are very useful. Larger genera have been divided into subgenera to facilitate identification.
8. The arrangement of taxa is based on overall **natural affinities** decided on the basis of morphological features, which can be easily studied with the naked eye or with a hand lens.
9. Although a few important characters have been chosen to name a few groups, the grouping itself is based on a **combination of characters**, rather than any single character in the majority of cases. Thus although *Delphinium* has fused petals, it has been kept in **Ranunculaceae** along with the related genera and placed in Polypetalae. Similarly, some gamopetalous genera of **Cucurbitaceae** are retained along with the polypetalous ones and placed in Polypetalae.
10. Heteromerae is rightly placed before Bicarpellatae.

Demerits

The system being pre-Darwinian in approach suffers from the following drawbacks:

1. The system does not incorporate phylogeny, although it was published after Darwin published his evolutionary theory.
2. **Gymnosperms** are placed between Dicotyledons and Monocotyledons, whereas their proper position is before the former, as they form a group independent from angiosperms.
3. **Monochlamydeae** is an unnatural assemblage of taxa, which belong elsewhere. The creation of this group has resulted in the separation of many closely related families. **Caryophyllaceae, Illecebraceae** and **Chenopodiaceae** are closely related families to the extent that they are

placed in the same order in all major contemporary classifications. Several authors including Takhtajan (1987) merge Illecebraceae with Caryophyllaceae. In Bentham and Hooker's system, however, Caryophyllaceae are placed in **Polypetalae**, and the other two in **Monochlamydeae**. Similarly, **Podostemaceae**, which are placed in a separate series Multiovulatae aquaticae, better belong to the order Rosales (Cronquist, 1988). **Chloranthaceae** placed by Bentham and Hooker under Microembryeae and Laurineae placed under Daphanales are closely allied to the order **Magnoliales** (Ranales *s. I.*) and are thus placed in the same subclass Magnoliidae by Cronquist (1988).

4. In Monocotyledons, **Liliaceae** and **Amaryllidaceae** are generally regarded as closely related and often included in the same order, some authors, including Cronquist, merging Amaryllidaceae with Liliaceae. In this system they are placed under different series, Amaryllidaceae under **Epigynae** and Liliaceae under **Coronarieae**.

5. **Unisexuales** is a loose assemblage of diverse families, which share only one major character, i.e., unisexual flowers. Cronquist (1988) separates these families under two distinct subclasses **Hamamelidae** and **Rosidae** and Takhtajan (1987) under **Hamamelididae** and **Dilleniidae**.

6. Bentham and Hooker did not know the affinities of the families placed under Ordines anomali, and the families were tentatively grouped together. Cronquist (1988) and Takhtajan (1987) place **Ceratophyllaceae** under subclass **Magnoliidae** and the other three under **Dilleniidae**.

7. Many large families, e.g. **Urticaceae**, **Euphorbiaceae** and **Saxifragaceae**, are unnatural assemblages and represent **polyphyletic** groups. These have rightly been split by subsequent authors into smaller, natural and **monophyletic** families.

8. **Orchidaceae** is an advanced family with inferior ovary and zygomorphic flowers, but the family is placed towards the beginning of Monocotyledons.

9. In Gamopetalae, Inferae with an inferior ovary is placed before the other two orders having a superior ovary. The inferior ovary is now considered to have been derived from a superior ovary.

ENGLER AND PRANTL

This is a system of classification of the whole plant kingdom, proposed jointly by two German botanists: Adolph Engler (1844-1930) (Fig. 7.3) and Karl A.E. Prantl (1849-1893). The classification was published in a monumental work *Die Naturlichen pflanzenfamilien* in 23 volumes (1887-1915). Engler was Professor of Botany at the University of Berlin and later Director, Berlin Botanic Garden. The system provided classification and description down to the **genus level**, incorporating information on morphology, anatomy and geography.

Fig. 7.3 Adolph Engler (1844-1930), the famous German botanist who produced the most comprehensive classification of the plant kingdom along with K. Prantl in a 23-volume work *Die Naturlichen pflanzenfamilien* (1887-1915) (reproduced with permission from Royal Botanic Gardens, Kew).

The system is commonly known under Engler's name, who first published classification up to the family level under the title *Syllabus der pflanzenfamilien* in 1892. This scheme was constantly revised by Engler and continued by his followers after his death, the latest 12th edition, edited by H. Melchior appearing in 2 volumes, 1954 and 1964. In this last edition, however, dicots were placed before monocots. Engler also initiated an ambitious plan of providing **taxonomic monographs** of various **families** up to species level under the title *Das pflanzenreich*. Between 1900 and 1953, 107 volumes were published covering 78 families of seed plants and one family (Sphagnaceae) of mosses.

This system often considered the beginning in **phylogenetic** schemes, was not strictly phylogenetic in the modern sense. It was an arrangement of linear sequence starting with the simplest groups and arranged in the order of **progressing complexity**. In doing so, unfortunately, Engler misread angiosperms, where in many groups, the **simplicity** is a result of **evolutionary reduction**. The system, however, had significant improvements over Bentham and Hooker: **Gymnosperms** were placed before angiosperms, group

Monochlamydeae was abolished and its members distributed along with their polypetalous relatives, and many large unnatural families were split into smaller natural families. The placement of monocots before dicots, another change made by this system did not, however, get subsequent support. The placement of the so-called group **Amentiferae** comprising families Betulaceae, Fagaceae, Juglandaceae, etc. in the beginning of dicots, also did not find much subsequent support. The system (Table 7.2) became very popular, like that of Bentham of Hooker, due to its comprehensive treatment and is still being followed in many herbaria of the world. Some recent floras including *Flora Europaea* (1964-1980) follow this system.

Table 7.2: Outline of the system of classification of Engler and Prantl

Plant Kingdom
>   **Division 1**
>   } Thallophytes
>   **Division 11**

Division 12. Embryophyta Asiphonogama

>   Subdivision 1. Bryophyta
>   Subdivision 2. Pteridophyta

Division 13. Embryophyta Siphonogama

>   Subdivision 1. Gymnospermae
>   Subdivision 2. Angiospermae
>   Class 1. Monocotyledoneae—*11 orders, 45 families*
>   Class 2. Dicotyledoneae—*44 orders, 258 families*
>   Subclass 1. Archichlamydeae (petals absent
>   or free)—*33 orders, 201 families*
>   Subclass 2. Metachlamydeae (petals united)—
>   *11 orders, 57 families*

In this scheme of classification the plant kingdom was divided into 13 divisions (in the 11th edition of *Syllabus der pflanzenfamilien* published in 1936, 14 divisions and in the 12th edition edited by Melchior 17 divisions were recognised), of which the first 11 dealt with Thallophytes, the 12th **Embryophyta Asiphonogama** (embryo formed, no pollen tube) included Bryophytes and Pteridophytes, while the 13th and last division **Embryophyta Siphonogama** (embryo formed, pollen tube developing) included seed plants.

Merits

The classification of Engler and Prantl has the following improvements over that of Bentham and Hooker (Table 7.3).

1. This was the first major system to incorporate the ideas of **organic evolution**, and the first major step towards phylogenetic systems of classification.

Table 7.3: Comparison of classification system of Bentham and Hooker with that of Engler and Prantl

Bentham and Hooker	Engler and Prantl
1. Published in *Genera Plantarum* in 2 volumes (1862-83).	1. Published in *Die Naturlichen pflanzenfamilien* in 23 volumes (1887-1915).
2. Includes only seed plants.	2. Includes the whole plant kingdom.
3. Gymnosperms placed in between Dicotyledons and Monocotyledons.	3. Gymnosperms separated and placed before the angiosperms.
4. Dicotyledons placed before Monocotyledons.	4. Dicotyledons placed after Monocotyledons.
5. Dicotyledons divided into 3 subclasses: Polypetalae, Gamopetalae and Monochlamydeae.	5. Dicotyledons divided into 2 subclasses: Archichlamydeae and Metachlamydeae.
6. Subclasses are further subdivided into series, cohorts (representing orders) and natural orders (representing families).	6. Subclasses are further subdivided into orders and families, series not recognised.
7. Monocotyledons include 7 series and 34 natural orders.	7. Monocotyledons include 11 orders and 45 families.
8. Pre-Darwinian in concept.	8. Post-Drawinian in concept.
9. Dicotyledons start with Ranales having bisexual flowers.	9. Dicotyledons start with Verticillatae with unisexual flowers.
10. Monocotyledons start with Microspermae including Orchidaceae.	10. Monocotyledons start with Pandanales. Microspermae are placed towards the end of Monocotyledons.
11. Closely related families Caryophyllaceae, Illecebraceae and Chenopodiaceae are kept apart, the first under Polypetalae and the other two under Monochlamydeae.	11. Family Illecebraceae is merged with Caryophyllaceae. Chenopodiaceae and Caryophyllaceae placed in the same order, Centrospermae.
12. Closely related families Amaryllidaceae and Liliaceae placed in separate series Epigynae and Coronarieae respectively.	12. Liliaceae and Amaryllidaceae placed in the same order, Liliiflorae.
13. Many larger families, e.g., Urticaceae, Saxifragaceae and Euphorbiaceae are unnatural heterogeneous groups.	13. Several larger families of Bentham and Hooker split into smaller homogeneous families. Urticaceae split into Urticaceae, Ulmaceae and Moraceae.

2. The classification covers the entire plant kingdom and provides description and identification keys down to the level of **family** (in *Syllabus der pflanzenfamilien*), **genus** (in *Die Naturlichen pflanzenfamilien*) and even **species** for large number of families (in *Das pflanzenreich*). Valuable illustrations and information on anatomy and geography are also provided.

3. **Gymnosperms** are separated and placed before angiosperms.

4. Many large unnatural families of Bentham and Hooker have been split into smaller and natural families. The family **Urticaceae** is thus split into **Urticaceae, Ulmaceae** and **Moraceae**.

5. Abolition of **Monochlamydeae** has resulted in bringing together several closely related families. Family **Illecebraceae** is merged with **Caryophyllaceae**. **Chenopodiaceae** and Caryophyllaceae are placed in the same order, **Centrospermae.**

6. **Compositae** in dicots and **Orchidaceae** in monocots are advanced families with inferior ovary, zygomorphic and complex flowers. These are rightly placed towards the end of dicots and monocots respectively.

7. Consideration of gamopetalous condition as advanced over polypetalous condition is in line with current phyletic views.

8. The classification being very thorough has been widely used in textbooks, Floras and herbaria around the world.

9. The terms **cohort** and **natural order** have been replaced by the appropriate terms order and family respectively.

Demerits

With better understanding of the phylogenetic concepts in recent years, many drawbacks of the system of Engler and Prantl have come to light. These primarily result from the fact that they applied the concept of 'simplicity representing primitiveness' even to the angiosperms, where evolutionary reduction is a major phenomenon, not commonly seen in the lower groups. The major drawbacks of the system include:

1. The system is not a phylogenetic one in the modern sense. Many ideas of Engler are now outdated.

2. Monocotyledons are placed before Dicotyledons. It is now widely agreed that Monocotyledons evolved from the primitive Dicotyledons.

3. The so-called **Amentiferae** including such families as Betulaceae, Juglandaceae and Fagaceae with reduced unisexual flowers, having few floral members and borne in catkins, were considered primitive. It has been established from studies on wood anatomy, palynology and floral anatomy that Amentiferae is an advanced group. The simplicity of flowers is due to **evolutionary reduction** and not primitiveness.

4. **Dichlamydeous** forms (distinct calyx and corolla) were considered to have evolved from the **monochlamydeous** forms (single whorl of perianth). This view is not tenable.

5. Angiosperms were considered a **polyphyletic** group. Most of the recent evidence points towards the monophyletic origin of angiosperms.

6. **Araceae** in Monocotyledons are now believed to have evolved from **Liliaceae**. In this classification Araceae are included in the order **Spathiflorae** which is placed before Liliiflorae including family Liliaceae.

7. **Helobieae** (including families Alismaceae, Butomaceae and Potamogetonaceae) is a primitive group, but in this classification placed after **Pandanales**, which is a relatively advanced group.

8. Derivation of **free central** placentation from **parietal** placentation, and of the latter from **axile** placentation is contrary to the evidence from floral anatomy. Free central placentation is now believed to have evolved from axile placentation through the disappearance of septa.

9. **Ranales** (in the broader sense—*s. l.*) are now considered a primitive group with bisexual flowers, spirally arranged floral parts and numerous floral members. In this classification they are placed much lower down, after Amentiferae.

JOHN HUTCHINSON

John Hutchinson (1884-1972) was a British botanist associated with the Royal Botanic Gardens, Kew, England who also served as keeper of Kew herbarium for many years (Fig. 7.4). He first proposed his classification of angiosperms in his book *The Families of Flowering Plants*, the first volume on Dicotyledons appearing in 1926 and the second on Monocotyledons in 1934. The classification was revised periodically, with the 2nd edition in 1959 and the 3rd in 1973, one year after his death.

Fig. 7.4 John Hutchinson (1884-1972), the British botanist, who worked as keeper of Kew Herbarium and published classification of angiosperms as *The Families of Flowering Plants* (1973), as also *The Genera of Flowering Plants* (reproduced with permission from Royal Botanic Gardens, Kew).

In addition to presenting his system of classification for angiosperms, Hutchinson also published such valuable works as *Flora of West Tropical Africa* (1927-29), *Common Wild Flowers* (1945), *A Botanist in South Africa* (1946), *Evolution and Classification of Rhododendrons* (1946), *British Flowering Plants* (1948), *More Common Wild Flowers* (1948), *Uncommon Wild Flowers* (1950), *British Wild Flowers* (1955), *Evolution and Phylogeny of Flowering Plants* (1969) and *Key to the Families of Flowering Plants of the World* (1968). He also embarked upon an ambitious plan of revising *Genera plantarum* of Bentham and Hooker under the title *The Genera of Flowering Plants*. Unfortunately he could complete only 2 volumes of this work, published in 1964 and 1967, the project cut short by his demise.

The classification system of Hutchinson dealt only with the flowering plants, included under **Phylum Angiospermae** as distinct from **Phylum Gymnospermae**. The classification was based on **24 principles** including General principles, Relating to General Habit, Relating to General Structure of Flowering Plants and those Relating to Flowers and Fruits. These **principles** are outlined below:

Other things being equal, it may be stated that:

1. Evolution is both upwards (sympetaly, epigyny) and downwards (apetaly, unisexuality).
2. Evolution does not necessarily involve all the organs of a plant at the same time; one organ or set of organs may be advancing while the other may be stationary or retrograding.
3. Evolution has generally been consistent and when a particular progression or retrogression has set in, it is persisted into the end of the phylum.

Relating to General Habit of Plants

4. In certain groups, trees and shrubs are probably more primitive than herbs.
5. Trees and shrubs are older than climbers, the latter habit having been acquired through a particular environment.
6. Perennials are older than biennials and annuals have been derived from them.
7. Aquatic Phanerogams are as a rule more recent than terrestrial (at any rate in the members of the same family or genus), and the same may be said of epiphytes, saprophytes and parasites.

Relating to General Structure of Flowering Plants

8. Plants with collateral vascular bundles arranged in a cylinder (Dicotyledons) are more primitive in origin than those with scattered bundles (Monocotyledons), though it does not necessarily follow that the latter have been directly derived from the former.
9. Spiral arrangement of leaves on the stem and of floral leaves precedes that of opposite and whorled types.
10. As a rule simple leaves precede compound leaves.

Relating to Flowers and Fruits of Plants

11. Bisexual precede unisexual flowers, and the dioecious is probably more recent than the monoecious condition.
12. Solitary flower is more primitive than the inflorescence.
13. Spirally imbricate floral parts are more primitive than whorled and valvate.
14. Many-parted flowers (polymerous) precede, and the type with few parts (oligomerous) follow from it, being accompanied by progressive sterilisation of reproductive parts.
15. Petaliferous flowers precede apetalous ones, the latter being the result of reduction.
16. Free petals (polypetaly) are more primitive than connate petals (sympetaly).
17. Actinomorphy of flower is an earlier type than zygomorphy.
18. Hypogyny is the primitive structure, and from it perigyny and epigyny were derived later.
19. Free carpels (apocarpy) are more primitive and from them connate carpels resulted.
20. Many carpels (polycarpy) precede few carpels (oligocarpy).
21. The endospermic seed with small embryo is primitive and the non-endospermic seed more recent.
22. In primitive flowers there are many stamens, in more advanced flowers few stamens.
23. Separate stamens precede connate stamens.
24. Aggregate fruits are more recent than single fruits, and as a rule the capsule precedes the drupe or berry.

Following Bessey, Hutchinson considered flowering plants to be **monophyletic**, having evolved from the hypothetical cycadeoid ancestral group which he gave the name of **Proangiosperms**. He recognised a number of smaller groups, bound together by a combination of characters. He established **Magnoliales** as an order distinct from **Ranales**, as he considered them to have evolved on parallel lines.

Hutchinson regarded **Magnoliaceae** as the most primitive family of the living angiosperms. He considered Dicotyledones to be more primitive and placed them (Table 7.4) before Monocotyledones, giving them a rank of **Subphylum**. The groups Polypetalae, Gamopetalae and Monochlamydeae were totally abolished; instead Hutchinson recognised two evolutionary lines: division **Lignosae** (fundamentally woody group) and **division Herbaceae** (fundamentally herbaceous group) within Dicotyledones, the former starting with **Magnoliaceae** and ending with **Verbenaceae**.

The Herbaceae started with **Paeoniaceae** and ended with **Lamiaceae**. Within Monocotyledones he recognised three evolutionary lines: division **Calyciferae** (calyx bearers), division **Corolliferae** (corolla-bearers) and division **Glumiflorae** (glume-bearers). A total of 411 families are recognised, 342 in Dicotyledones and 69 in Monocotyledones. Lignosae includes 54 orders, Herbaceae 29, Calyciferae 12, Corolliferae 14 and Glumiflorae 3. A diagram (appropriately **phylogram**) showing phylogeny and evolution within Dicotyledones is presented in Fig. 7.5.

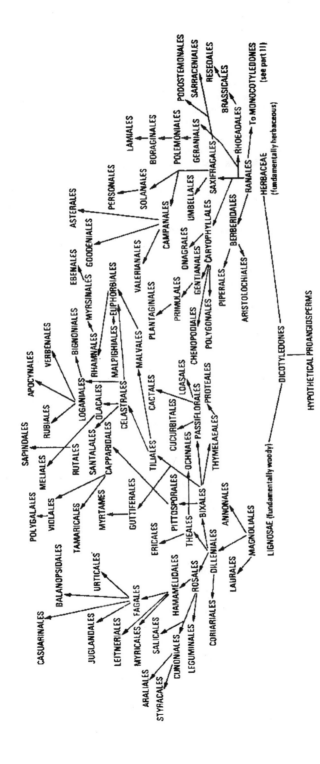

Fig. 7.5 Hutchinson's diagram (phylogram) showing phylogeny and relationships of orders of Dicotyledones as presented in his 1973 classification.

Table 7.4: Outline of the system of classification of flowering plants presented by Hutchinson in 3rd edition of *The Families of Flowering Plants* (1973).

Phylum I. Gymnospermae
Phylum II. Angiospermae
  Subphylum I. Dicotyledones
    Division I. Lignosae—*54 orders*
      Order 1. Magnoliales (first family Magnoliaceae)
         . . . .

      Order 54. Verbenales (last family Verbenaceae)
    Division II. Herbaceae—*28 orders*
      Order 55. Ranales (first family Paeoniaceae)
         . . . .

      Order 82. Lamiales (last family Lamiaceae)
  Subphylum II. Monocotyledones
    Division I. Calyciferae—*12 orders*
      Order 83. Butomales (first family Butomaceae)
         . . . .

      Order 94. Zingiberales (last family Marantaceae)
    Division II. Corolliferae—*14 orders*
      Order 95. Liliales (first family Liliaceae)
         . . . .

      Order 108. Orchidales (family Orchidaceae only)
  Division III. Glumiflorae—*3 orders*
      Order 109. Juncales (first family Juncaceae)
         . . . .

      Order 111. Graminales (family Poaceae only)

Whereas Hutchinson considered the woody habit to be primitive in dicots, in monocots he considered the herbaceous habit to be primitive, and the woody forms derived from the herbaceous forms. He considers Monocotyledones also to be a monophyletic group derived from Ranales, Butomales having a link with Helleboraceae and Alismatales with Ranunculaceae. The presence of endosperm in seeds of Ranunculaceae and its absence from Butomaceae and Alismataceae, otherwise considered closer, is explained by Hutchinson to be the result of aquatic habit in the last two. A diagram (**phylogram**) showing the probable phylogeny of various orders in Monocotyledones is presented in Fig. 7.6.

MERITS

The system of Hutchinson, being based on a number of sound phylogenetic principles, and studies of a large number of plants at his disposal at Kew, shows the following improvements over earlier systems:

1. The system is more **phylogenetic** than that of Engler and Prantl, as it is based on phylogenetic principles generally recognised by most authors.
2. The treatment of **Magnoliales** as the starting point in the evolutionary series of Dicotyledones is in agreement with prevalent views.

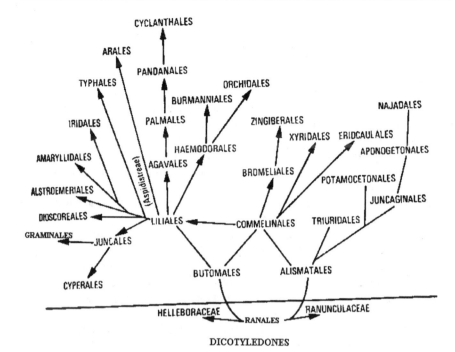

CYCLANTHALES

ARALES

PANDANALES

TYPHALES

ORCHIDALES

BURMANNIALES

NAJADALES

IRIDALES   PALMALES

ZINGIBERALES

XYRIDALES   ERIOCAULALES

HAEMODORALES

APONOGETONALES

AMARYLLIDALES

(Aspidistreae)   AGAVALES

BROMELIALES

POTAMOCETONALES

ALSTROEMERIALES

JUNCAGINALES

DIOSCOREALES

LILIALES

COMMELINALES   TRIURIDALES

GRAMINALES

JUNCALES

BUTOMALES   ALISMATALES

CYPERALES

HELLEBORACEAE   RANUNCULACEAE

RANALES

DICOTYLEDONES

Fig. 7.6 Hutchinson's diagram (phylogram) showing probable phylogeny and relationship of orders within Monocotyledones.

3. The abolition of Polypetalae, Gamopetalae, Monochlamydeae, Archichlamydeae and Metachlamydeae and rearrangement of taxa on the **combination of characters** and not one or a few characters as in earlier systems is more logical.

4. Many large unnatural families have been split into smaller natural ones. **Euphorbiaceae** of Bentham and Hooker is split into Euphorbiaceae, Ricinaceae and Buxaceae. The family **Urticaceae** is similarly split into Urticaceae, Moraceae, Ulmaceae and Cannabinaceae.

5. Standards of description are very high. Useful keys are provided for the identification of families.

6. **Phylograms** for dicots and monocots are more superior than the Besseyan cactus.

7. The classification of **Monocotyledones** is sounder and generally appreciated; even keys to the identification of **genera** have been provided.

8. The derivation of Monocotyledones from Dicotyledones is widely agreed.

9. The placement of **Alismatales** towards the beginning of Monocotyledones finds general acceptance.

10. Detailed classification up to generic level, together with identification keys and description has been provided for some families in the two volumes of *The Genera of Flowering Plants*.

Demerits

The classification of Hutchinson has largely been ignored, as it mostly did not proceed beyond family level, and gave much importance to the habit. The major drawbacks of the system are listed below:

1. The system is not useful for practical identification, arrangement in Floras and herbaria, as it does not proceed beyond the family level in the greater majority of taxa.
2. The division of Dicotyledones into **Lignosae** and **Herbaceae** is most artificial and has resulted in separation of closely related families **Araliaceae** and **Apiaceae**, in Lignosae and Herbaceae respectively. Lamiaceae and Verbenaceae are similarly very closely related and often placed in the same order in contemporary systems of classification. Hutchinson, on the basis of habit, separated them under distinct orders and even separate divisions Herbaceae and Lignosae respectively.
3. Hutchinson did not provide a full explanation for the majority of his evolutionary concepts.
4. He derives angiosperms from **proangiosperms**, but does not provide information about the nature of this hypothetical ancestral group.
5. Although he has split several large unnatural families into natural units, in some cases he has even split some families which were already natural monophyletic groups. The family **Ranunculaceae** has been split into **Ranunculaceae** and **Helleboraceae** on the basis of achene and follicle fruit, respectively. Studies on the floral anatomy have shown that evolutionary stages in the reduction of ovule number can be seen in the genera of Helleboraceae, and many genera of Ranunculaceae show traces which would have gone to now aborted ovules. Thus the Ranunculaceae of Bentham and Hooker represents a monophyletic group and need not be split.
6. The family **Calycanthaceae** is related to Laurales, but placed here in Rosales.
7. Hutchinson regards Magnoliaceae as the most primitive family of living Dicotyledones, but most contemporary authors consider vesselless **Winteraceae** to be the most primitive.

ARMEN TAKHTAJAN

Armen Takhtajan (1910-97) was a leading Russian plant taxonomist (Fig. 7.7) and chief of the Department of higher plants in V.L. Komarov

Fig. 7.7   Armen Takhtajan (1910-1997), leading Soviet authority on phytogeography and classification of flowering plants. Published last version of his classification in 1997, incorporating several modifications.

Botanic Institute, USSR Academy of Sciences, Leningrad (now named St. Petersburg). He was an international authority on phytogeography, origin and phylogeny of flowering plants. He was the President of the 12th International Botanical Congress held in Leningrad in 1975.

His classification was first published in 1954 in Russian, but came to be known outside the Soviet Union only after its English translation *Origin of Angiospermous Plants* was published in 1958. The system was elaborated in *Die evolution der angiospermen* (1959), and *systema et phylogenia magnoliophytorum* (1966), both in Russian. The classification became popular with the English translation of the latter as *Flowering Plants—Origin and Dispersal* by C. Jeffrey in 1969. The classification was published in revised form in *Botanical Review* in 1980. A more elaborate revision of this classification appeared in the Russian work *Sistema Magnoliophytov* (Latin facsimile *Systema Magnoliophytorum*) in 1987. Between 1980 and 1987 he proposed smaller revisions in 1983 (revision of dicots only in Metcalfe and Chalk: *Anatomy of Dicotyledons*, vol. 2) and 1986 (Takhtajan: *Floristic Regions of the World*). His final comprehensive system of classification was published in 1997 (*Diversity and Classification of Flowering Plants*). Earlier Takhtajan, along with Cronquist and Zimmerman, had also provided a broad classification of **Embryobionta** (1966).

Takhtajan, who has provided a classification of angiosperms up to the family level, belonged to the Besseyan School and was strongly influenced by Hutchinson, Hallier and the other more progressive German workers.

Table 7.5: Outline of the system of classification of Angiosperms proposed by Takhtajan in 1997. Subclasses with * did not exist in 1987 classification.

Division. Magnoliophyta—*2 classes, 17 subclasses, 72 superorders, 233 orders, 592 families (2 classes, 12 subclasses, 53 superorders, 166 orders, 533 families in 1987 classification); estimated genera—13,000,—species 2,50,000*

    Class 1. Magnoliopsida (Dicotyledons)—*11 subclasses, 56 superorders, 175 orders, 459 families (8 subclasses, 37 superorders, 128 orders, 429 families in 1987 classification); estimated genera—10,000, species—1,90,000*

        Subclass 1. Magnoliidae
                2. Nymphaeidae*
                3. Nelumbonidae*
                4. Ranunculidae
                5. Caryophyllidae
                6. Hamamelididae
                7. Dilleniidae
                8. Rosidae
                9. Cornidae*
             10. Asteridae
             11. Lamiidae

    Class 2. Liliopsida (Monocotyledons)—*6 subclasses, 16 superorders, 58 orders and 133 families (4 subclasses, 16 superorders, 38 orders, 104 families in 1987 classification); estimated genera—3,000, species—60,000*

        Subclass 1. Liliidae
                2. Commelinidae*
                3. Arecidae
                4. Alismatidae
                5. Triurididae
                6. Aridae*

As against the classification proposed in 1980, the revision proposed in 1987 had one subclass each added to Magnoliopsida (only former superorder Asteranae—now split into Asteranae and Campanulanae—retained in Asteridae, all remaining superorders placed in a new subclass **Lamiidae**) and Liliopsida (superorder Triuridanae separated from Liliidae into a new subclass **Triuridinae**). Also 17 superorders, 56 orders and 96 families were added to Magnoliopsida and 8 superorders, 17 orders and 27 families to Liliopsida. In his 1997 revision he added three new subclasses **Nymphaeidae, Nelumbonidae** (separated from Magnoliidae) and **Cornidae** (separated from Rosidae) to Magnoliopsida and two **Commelinidae** (separated from Liliidae) and **Aridae** (separated from Arecidae). He further added 19 superorders, 47 orders and 30 families in Magnoliopsida and 20 orders and 29 families in Liliopsida.

An interesting aspect of the 1987 classification of Takhtajan was uncertainty about the placement of the family **Cynomoriaceae**. The single genus *Cynomorium* earlier placed in family Balanophoraceae (Hutchinson, 1973; Cronquist, 1988), was removed to the family Cynomoriaceae and placed next to Balanophoraceae under the order Balanophorales by Takhtajan (1980), Thorne (1983,1992) and Dahlgren (1983,1989). In his 1987 classification,

Takhtajan had placed this family under the order **Cynomoriales**, but not being certain about its affinities, inserted this order tentatively towards the end of Rosidae. In his 1997 classification he brought Cynomoriales under Magnoliidae within superorder **Balanophoranae**.

A major departure of Takhtajan from earlier versions is the recognition of **Commelinidae** as a distinct subclass in agreement with the position taken by Cronquist. Takhtajan, however, unlike Cronquist placed **Liliidae** at the beginning of Liliopsida, while the Alismatidae are placed higher up after Arecidae. Like other phylogenetic systems of classification, the presumed relationship of various subclasses and superorders is indicated with the help of a **bubble diagram** (Fig. 7.8 for dicots; Fig. 7.9 for monocots)—more appropriately a **phylogram**—the size of each bubble or balloon representing the relative size of each group, the branching pattern the phylogenetic relationship, and the length of bubble its evolutionary advancement (**degree of apomorphy**).

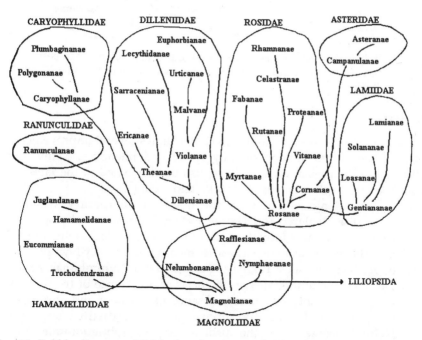

Fig. 7.8 Bubble diagram of Takhtajan showing probable relationship between different subclasses and superorders of dicotyledons (based on Takhtajan, 1987). 1997 classification does not include a bubble diagram.

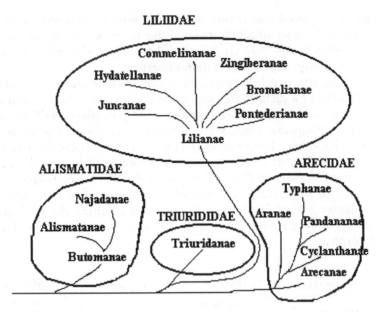

Fig. 7.9 Bubble diagram of Takhtajan showing probable relationship between different subclasses and superorders of monocotyledons (based on Takhtajan, 1987).

Merits

The latest classification of Takhtajan (1997) shows several improvements in light of recent information on phylogeny and phenetics. Many merits achieved in the earlier versions have also been retained in the latest revision. The major achievements of this system include:

1. A general agreement with the major contemporary systems of Cronquist, Dahlgren and Thorne and incorporation of phylogenetic as well as phenetic information for the delimitation of orders and families. The genus *Nelumbo* was earlier placed in the family Nymphaeaceae under Nymphaeales. Takhtajan separated it to **Nelumbonaceae** under the order **Nelumbonales** on the basis of occurrence of tricolpate pollen grains, embryo structure, absence of laticifers and chromosome morphology. He finally separated it to a distinct superorder **Nelumbonanae** under the distinct subclass **Nelumbonidae**. Thorne (1983, 1992) also follows the separation into Nelumbonales but under superorder Ranunculanae. Similarly, the genus *Eucommia* was earlier placed in the family Hamamelidaceae. Takhtajan removed it to the family **Eucommiaceae** under the order **Eucommiales** based on the presence of stipules, unilacunar nodes, unitegmic ovule and cellular endosperm, a separation followed by Cronquist (1988). Thorne (1983,

1992) gives it a rank of suborder under the order Hamamelidales. De Soo (1975) placed it under a separate subclass Eucommiidae. Similarly, the genus *Paeonia*, placed under the family Ranunculaceae in earlier classifications, was separated by Takhtajan to the family **Paeoniaceae** under the order **Paeoniales** on the basis of evidence from chromosomes (5 large chromosomes), floral anatomy (centrifugal stamens, many traces in sepals and petals, 5 in carpels), and embryology (unique embryogeny with coenocytic proembryo stage, reticulately pitted exine, large generative cell, thick fleshy carpels, broad stigmas, prominent lobed fleshy nectariferous disc surrounding the gynoecium). Thorne (1983, 1992) follows the separation of Paeoniaceae under the order Paeoniales.

2. The system is more **phylogenetic** than that of Hutchinson and other earlier authors and is based on now widely accepted phylogenetic principles.

3. The derivation of **Monocotyledons** from the terrestrial hypothetical extinct group of Magnoliidae (often called proangiosperms), is largely favoured, as also the view that **Alismatales** and **Nymphaeales** represent ancient side branches and have a common origin.

4. Abolition of artificial group names Polypetalae, Gamopetalae, Lignosae, Herbaceae etc. has resulted in more natural grouping of taxa. **Lamiaceae** and **Verbenaceae** are thus brought together under the order **Lamiales** (as against their separation under Lamiales and Verbenales and placement under separate groups Herbaceae and Lignosae respectively by Hutchinson). **Caryophyllaceae**, **Chenopodiaceae** and **Portulacaceae** have similarly been placed under the order Caryophyllales.

5. Nomenclature is in accordance with the International Code of Botanical Nomenclature, even up to the level of division.

6. Clifford (1977) from numerical studies has largely supported the division of **Monocotyledons** into subclasses.

7. The placement of Magnoliidae as the most primitive group of angiosperms, Dicotyledons before Monocotyledons, **Magnoliales** at the beginning of Magnoliopsida, finds general agreement with other authors.

8. Depiction of the putative relationships of major subclasses and superorders with the help of a **bubble diagram** is very useful. It gives some idea of the **relative size** of different groups, point of **cladistic divergence** and **degree of advancement** (apomorphy) reached. Larger groups are represented by larger bubbles, vertical length the degree of advancement, and the point of separation of a branch its cladistic divergence.

9. By splitting **Asteridae** into two subclasses: **Lamiidae** and **Asteridae**, a more rational distribution of sympetalous families has been achieved. Also **Cornanae**, which show affinities with the sympetalous families,

have been placed in an independent subclass **Cornidae** (earlier placed under Rosidae).

10. Removal of the order Urticales from Hamamelididae and its placement in an independent superorder **Urticanae**, between Malvanae and Euphorbianae under Dilleniidae is more appropriate. Dahlgren (1983) had pointed out affinities of Urticales with Malvales and Euphorbiales. The arrangement agrees with that of Thorne (1983,1992) also. Cronquist (1988), however, placed Urticales in Hamamelidae, Malvales in Dilleniidae and Euphorbiales in Rosidae.

11. The placement of **Dioncophyllaceae** in a separate order Dioncophyllales is in line with the opinion presented by Metcalfe and Chalk (1983), who on the basis of anatomical evidence, proposed that the family occupied an isolated taxonomic position. Earlier, the family had been included in the order Theales next to the family Ancistrocladaceae.

12. Nymphaeales whose position within the dicots has been a matter of debate have been placed in a distinct subclass **Nymphaeidae** under Magnoliopsida.

13. The ending **-anae**, earlier opposed in favour of **-florae** has now been accepted by G. Dahlgren (1989) and Thorne (1992) since the ending **-florae** restricts usage to angiosperms and is not universal in application.

Demerits

With the latest revision of his classification in 1997, Takhtajan attempted to remove deficiencies in earlier versions of his system. The critical appraisal of his latest version, in future, may bring out some further drawbacks. The following limitations of the system can be recorded:

1. The system, although very sound and highly phylogenetic, is not helpful for identification and for adoption in herbaria, as it provides classification only up to the family level. Also, keys to the identification of taxa are not provided.

2. Dahlgren (1983), Dahlgren et al. (1981) and Thorne (1983,1992) consider that the **angiosperms** deserve a **class** rank equivalent to the main groups of gymnosperms such as Pinopsida, Cycadopsida etc.

3. Clifford (1977) by numerical analysis has shown that **Arales** are closer to Liliales. Dahlgren (1983, 1989) placed Arales next to Liliiflorae. Clifford has also shown that **Gramineae** and **Cyperaceae** are not related to Liliidae and Commelinidae but Arecidae.

4. Although the system is based on data derived from all sources, in final judgement more weightage is given to **cladistic** information compared to phenetic information.

5. Ehrendorfer (1983) points out that **Hamamelididae** do not represent an ancient side branch of Magnoliidae, but are remnants of a transition from Magnoliidae to Dilleniidae-Rosidae-Asteridae.

6. Behnke (1977) and Behnke and Barthlott (1983) point out that Caryophyllales have PIII-type plastids whereas Polygonales and Plumbaginales have S-type plastids, and thus advocate their removal from Caryophyllidae to Rosidae, retaining only Caryophyllales in the subclass Caryophyllidae. Though not agreeing on their removal, Takhtajan (1987) partly incorporated Behnke's suggestion by placing all the three orders under separate superorders Caryophyllanae, Polygonanae and Plumbaginanae, but within the same subclass, Caryophyllidae.

7. Further splitting and increase in the number of families to 592 (533 in 1987) has resulted in a very narrow circumscription by creation of numerous **monotypic families** such as Pottingeriaceae, Barclayaceae, Hydrastidaceae, Nandinaceae, Griseliniaceae, Hypecoaceae etc., and numerous **oligotypic** ones such as Balanophoraceae, Sarraceniaceae, Peganaceae and Agrophyllaceae.

8. Most authors including Cronquist (1988) and Thorne (1983, 1992), regard the vesselless family **Winteraceae** as the most primitive among living angiosperms, but Takhtajan regarded **Degeneriaceae** as most primitive, considering Winteraceae an isolated group and placing it in a separate order Winterales.

ARTHUR CRONQUIST

Arthur Cronquist (1919-1992), a leading American taxonomist, associated with the New York Botanical Garden (Fig. 7.10), produced a broad classification of **Embryobionta** along with Takhtajan and Zimmerman (1966). He produced the first detailed classification of angiosperms in 1968 in his book *The Evolution and Classification of Flowering Plants*. The classification is conceptually similar to that of Takhtajan's system but differs in details. The classification was further elaborated in 1981 in his book *An Integrated System of Classification of Flowering Plants*. Some realignments in Dicotyledons were published in the *Nordic Journal of Botany* in 1983. The classification was finally revised in 1988.

The classification like that of Takhtajan, is based on evidence derived from all sources, although he gives more importance to morphology (Ehrendorfer, 1983). Following Takhtajan, the angiosperms are given the name Magnoliophyta and divided into Magnoliopsida (dicots) and Liliopsida (monocots). He includes only six subclasses in dicots and recognises five in monocots. In dicotyledons, the Ranunculidae of Takhtajan are merged with Magnoliidae and Lamiidae are not given a separate rank at subclass level but retained in Asteridae. In monocotyledons Zingiberidae are treated separate from Liliidae and Triuridales kept under Alismatidae. As a major departure from the systems of Takhtajan, Dahlgren and Thorne, no superorders are recognised; the subclasses are divided into orders directly.

Fig. 7.10 Arthur Cronquist (1919-1992), leading American Plant taxonomist who published a 2nd edition of his *The Evolution and Classification of Flowering Plants* in 1988. His classification is similar to that of Takhtajan in general outline (photograph courtesy Allen Rokach, The New York Botanical Garden Bronx, New York).

Also, as against 233 orders and 592 families recognised by Takhtajan, Cronquist recognises 83 orders and 386 families. Cronquist agrees with Thorne in keeping the family **Winteraceae** (and not Degeneriaceae as done by Takhtajan) at the beginning of dicotyledons, and included along with Degeneriaceae, Magnoliaceae, Annonaceae etc. in the same order **Magnoliales**. Paeoniaceae, unlike other contemporary authors, are not separated by Cronquist into a distinct order Paeoniales, but instead shifted to order Dilleniales under Dilleniidae.

Another significant departure from Takhtajan's system is the merger of Amaryllidaceae with Liliaceae, under the order Liliales. Takhtajan places these two families in separate orders Amaryllidales and Liliales respectively. Unlike most recent authors, Cronquist believed in the aquatic origin of monocotyledons, from primitive vesselless ancestors resembling present-day Nymphaeales.

He died working on a manuscript for *Intermountain Flora* in the herbarium at Brigham Young University in Provo, Utah. His death on 22 March 1992 marked the end of a period in systematic botany when a handful of men and women dominated much of the intellectual thought, not only in taxonomy but in much of botany.

Arthur was born on 19 march 1919 in San Jose, California, and grew up in Portland, Oregon. For his doctoral work, the "big blonde Swede"—as some called him—went to the University of Minnesota, doing a revision of the North American species of *Erigeron*. He first went to NYBG in 1943 as an assistant curator.

He and Arthur Holmgren, at Utah State University, initiated a study of the Intermountain West in 1959, with Cronquist concentrating on the multivolumed, illustrated flora of the Pacific Northwest with C. Leo Hitchcock and Marion Ownbey. He worked on the latter flora into the early 1970s when his focus moved southward. As senior author of the *Intermountain Flora* he championed the effort to summarize the plants of that region in a second illustrated, several-volumed flora. At the same time, he worked on revisions of Henry Gleason's and his own *Manual of Vascular Plants of Northeastern United States and Adjacent Canada*, a second edition appearing in 1991.

As an educator, Cronquist wrote two of the finest botany textbooks of the 1960s. *Introductory Botany* and *Basic Botany* were widely used and translated into several languages. He taught numerous students and several today teach at universities throughout the nation.

By the late 1960s, Cronquist moved into the category of senior statesman for botany. He lectured widely, was actively sought as a consultant, and wrote numerous articles on evolution, phylogeny and botany in general. He played major roles in the leading national and international organisations, serving as president of several including the American Society for Plant Taxonomy.

Arthur's interest in flowering plant classification first made its appearance in a paper he published in 1957 while working in Belgium. In 1965 he travelled to (then)·Leningrad to talk to Armen Takhtajan, and the two began a long and productive relationship. While both men published independently, and basically disagreed on many things, the two were in concert on many aspects of angiosperm phylogeny. Both published several versions with Cronquist opting for fewer modifications. His book, *The Evolution and Classification of Flowering Plants* published in 1968 set the stage for his future efforts, culminating in *An Integrated System of Classification of Flowering Plants* in 1981. It would be this later work that would set the stage for a massive renewal of interest in angiosperm phylogeny for here was a thorough work that could be tested using modern systematic means.

In contrast to Takhtajan's system Nelumbonaceae are placed in Nymphaelaes (and not a separate order Nelumbonales). Typhales in Commeliniidae (and not Arecidae), and sympetalous families of dicotyledons placed in a large subclass Asteridae (and not three subclasses Asteridae, Cornidae and Lamiidae). Urticales are included along with wind-pollinated families under Hamamelidae (and not with its related orders Malvales and Euphorbiales), and Malvales and Euphorbiales are kept in separate sub-classes, Dilleniidae and Rosidae respectively (and not the same subclass Dilleniidae). Cronquist has provided a synoptic arrangement of taxa, facilitating the process of identification up to the family level. An outline of Cronquist's system is presented in Table 7.6. The system is widely used in the USA.

Table 7.6: Broad outline of the classification of angiosperms presented by Cronquist (1988).

Division. Magnoliophyta—*2 classes, 11 subclasses, 83 orders and 386 families; 219,300 species*
   Class 1. Magnoliopsida (Dicotyledons)—*6 subclasses, 64 orders, 320 families; 169,400 species*
      Subclass 1. Magnoliidae
            2. Hamamelidae
            3. Caryophyllidae
            4. Dilleniidae
            5. Rosidae
            6. Asteridae
   Class 2. Liliopsida (Monocotyledons)—*5 subclasses, 19 orders, 66 families; 49,900 species*
      Subclass 1. Alismatidae
            2. Arecidae
            3. Commelinidae
            4. Zingiberidae
            5. Liliidae

The relationships of various subclasses and orders (Fig. 7.11) are shown with the help of a phylogram which takes the form of a **bubble diagram**, like other contemporary systems of classification.

Merits

The classification of Cronquist is largely based on principles of phylogeny, which find acceptance with major contemporary authors. The system is merited with the following achievements over the previous systems of classification:

1. It shows general agreement with major contemporary systems of Takhtajan, Dahlgren and Thorne, and incorporates evidence from all sources in arrangement of various groups. *Paeonia* and *Nelumbo* are thus placed under **Paeoniaceae** and **Nelumbonaceae**, although the orders Paeoniales and Nelumbonales are not recognised. *Eucommia* is also kept in a separate family Eucommiaceae under a distinct order **Eucommiales**.

2. The revision of the classification in 1981 and 1988 was presented in comprehensive form, giving detailed information on phytochemistry, anatomy, ultrastructure and chromosomes besides morphology.

3. The text being in English has been readily adopted in books and floristic projects originating in the USA.

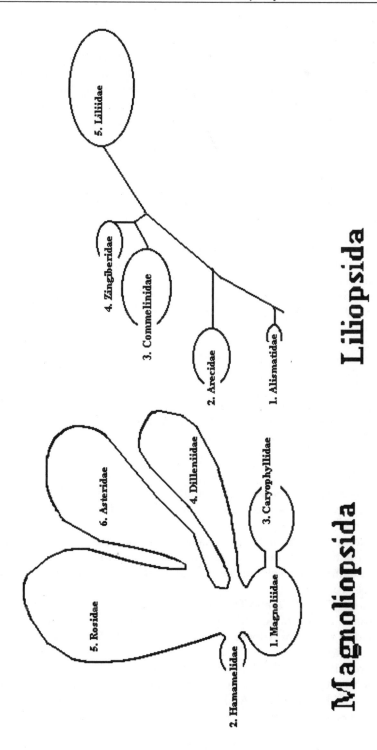

Fig. 7.11 Phylogram showing the relationship between various subclasses and orders as presented by Cronquist (based on Cronquist, 1988).

4. The system is highly **phylogenetic** and is based on now largely accepted phylogenetic principles.
5. The placement of **Winteraceae** at the beginning of dicotyledons is generally favoured by most authors including Ehrendorfer (1968), Gottsberger (1974) and Thorne (1976, 1992). The family has vesselless wood similar to gymnosperms, great similarity between micro- and megasporophylls, unifacial stamens and carpels, morphology similar to **pteridosperms**, high chromosome number suggesting long evolutionary history and less specialised beetle pollination compared to the genus *Magnolia*.
6. Abolition of artificial group names such as Polypetalae, Gamopetalae, Lignosae, Herbaceae etc. has resulted in more natural grouping of taxa. Verbenaceae and Lamiaceae are thus brought under the order **Lamiales**. Caryophyllaceae, Chenopodiaceae and Portulacaceae are similarly placed in the same order **Caryophyllales**.
7. **Nomenclature** is in accordance with the International Code of Botanical Nomenclature.
8. Placement of **Magnoliidae** as the most primitive group of angiosperms, dicotyledons before monocotyledons, Magnoliales at the beginning of Magnoliidae and Butomaceae at the beginning of Liliopsida, finds general agreement with other authors.
9. Compositae in dicotyledons and Orchidaceae in monocotyledons are generally regarded as advanced families, and are rightly placed towards the end of each group respectively.
10. The relationship of various groups has been depicted with diagrams, which provide valuable information on relative advancement, cladistic relationship and size of various subclasses.

Demerits

The system is becoming increasingly popular, especially in the USA, where many books are following this system. The following drawbacks, however, may be pointed out:

1. In spite of being a highly phylogenetic and popular in the USA, the system is not very useful for **identification** and adoption in herbaria since identification keys for genera, their distribution and description are not provided.
2. Dahlgren (1983, 1989) and Thorne (1980, 1983) considered angiosperms to deserve a **class** rank, and not that of a division.
3. **Asteridae** represent a loose assemblage of several diverse sympetalous families.
4. Clifford (1977) on the basis of numerical studies has shown that Typhales are better placed in Arecidae. Cronquist places Typhales in Commelinidae. Similarly **Arales** are not close to Arecales but Liliales. Gramineae and Cyperaceae have also been shown by Clifford to be

related to Arecidae and not Commelinidae. He also shows close relationship between Liliaceae and Commelinaceae, thus refuting their placement in different subclasses.

5. **Superorder** as a rank above the order is not recognised, thus showing a significant departure from contemporary systems of Takhtajan, Thorne and Dahlgren.

6. Ehrendorfer (1983) pointed out that **Hamamelidae** do not represent an ancient side-branch of a Magnoliidae but are remnants of a transition from Magnoliidae to Dilleniidae-Rosidae-Asteridae.

7. Behnke (1977)and Behnke and Barthlott (1983) advocate that **Polygonales** and **Plumbaginales** with S-type plastids should be removed to **Rosidae** and only Caryophyllales with PIII-type plastids retained in Caryophyllidae.

8. Urticales are placed in Hamamelidae together with wind-pollinated families, whereas they are close to Malvales and Euphorbiales (Dahlgren, 1983, 1989). Cronquist further separates Malvales in Dilleniidae and Euphorbiales in Rosidae.

9. Most recent authors do not believe in the aquatic ancestry of monocotyledons. Kosakai et al. (1970) have provided ample evidence to refute the aquatic ancestry of monocotyledons on the basis of study of primary xylem in the roots of *Nelumbo* (Nymphaeales). Cronquist believed that monocotyledons arose from vesselless ancestors resembling present-day Nymphaeales. Dahlgren et al. (1985) point out that Nymphaeales and Alismatales demonstrate a case of **multiple convergence**, and only a few characters (sulcate pollen grains and trimerous flowers) are due to shared ancestry. The presence of two cotyledons, S-type sieve tube plastids, occurrence of ellagic acid and perispermous seeds in Nymphaeales argue strongly against their position as a starting point of monocotyledons, and none of these attributes occur in Alismatales.

10. Metcalfe and Chalk (1983) on the basis of a unique combination of anatomical features suggested that family **Dioncophyllaceae** should occupy an isolated taxonomic position, but it was placed by Cronquist in order Violales before family Ancistrocladaceae.

11. Cronquist (1988) recognised Physenaceae as a family under Order Urticales, but was not sure about its exact placement.

8

Taxonomic Evidence

Over the last few decades the affinities between plant groups have been redefined as more and more information is accumulated from various sources. Newer approaches in recent years have been a) increasing reliance on phytochemical information (**Chemotaxonomy**), b) studies on ultrastructure and micromorphology, c) statistical analysis of available data without much a priori weighting and providing a synthesis of all information (**Taxometrics**), and d) analysis of phylogenetic data to construct phylogenetic relationship diagrams (**Cladistics**). The aforesaid disciplines constitute the major **modern trends in taxonomy**. Data continues to flow from different disciplines, so that the process of analysis and synthesis is an ongoing activity. **Taxonomy is as such a field of unending synthesis**. The following disciplines have contributed to a greater or lesser extent to a better understanding of taxonomic affinities between plants.

I. MORPHOLOGY

Morphology has been the major criterion for classification over the last so many centuries. The initial classifications were based on gross morphological characters. During the last two centuries more and more microscopic characters of morphology were incorporated. Although floral morphology has been the major material for classifications, other morphological characters have also contributed in specific groups of plants.

Habit

Life-forms though of little significance to taxonomy, allow a means of estimating adaptiveness and ecological adjustment to the habitat. In *Pinus* bark characters are used for identification of species. Woody and herbaceous characters have been the primary basis of recognition of Lignosae and Herbaceae series within dicots by Hutchinson (1926, 1973).

Underground Parts

Rhizome characteristics are important for identification of various species of the genus *Iris*. Similarly bulb structure (whether bulbs are clustered on rootstock or not) is an important taxonomic criterion in the genus *Allium*. Davis (1960) has divided Turkish species of the subgenus *Ranunculus* of genus *Ranunculus* based on rootstock and habit.

Leaves

Leaves are important for identification in palms, *Salix* and *Populus*. The genus *Azedirachta* has been separated from *Melia* among other features by the presence of unipinnate leaves as against bipinnate in the latter. Similarly, the genus *Sorbus* has been separated from *Pyrus*, and genus *Sorbaria* separated from *Spiraea* on the basis of pinnate leaves. Stipules are an important source for identification in *Viola* and *Salix*. Leaf venation is important for the identification of the species in *Ulmus* and *Tilia*. Interpetiolar stipules are useful for identification within family Rubiaceae.

Flowers

Floral characters are extensively used in delimitation of taxa. These may include the calyx (Lamiaceae), corolla (Fabaceae, *Corydalis*), stamens (Lamiaceae, Mimosaceae), or carpels (Caryophyllaceae). A gynobasic style is characteristic of Lamiaceae. Similarly the gynostegium characterises Asclepiadaceae.

Fruits

Fruit characteristics are very widely used in identification. Coode (1967) used only fruit characteristics in delimitation of species of the genus *Valerianella*. Singh et al. (1972) used fruit morphology in identification of Indian genera of Compositae (Liguliflorae). The number of capsule valves is used in segregating genera in family Caryophyllaceae (*Melandrium, Silene, Cerastium*). Seed characters are valuable identification features in the genus *Veronica*.

II. ANATOMY

Anatomical features have played an increasingly important role in elucidation of phylogenetic relationships. Anatomical work of taxonomic significance was largely undertaken by Bailey and his students. Carlquist (1996) has discussed the trends of xylem evolution especially in the context of primitive angiosperms.

Wood Anatomy

Studies on wood anatomy have contributed largely in arriving at the conclusion that Amentiferae constitute a relatively advanced group, and that Gnetales are not ancestral to angiosperms. Bailey (1944) concluded that vessels in angiosperms arose from tracheids with scalariform pitting, whereas in Gnetales they arose from tracheids with circular pitting, thus suggesting an independent origin of vessels in these two groups. Demonstration of vesselless angiosperms (Winteraceae, Trochodendraceae), also having other primitive features, has led to the conclusion that angiosperm ancestors were vesselless. The separation of *Paeonia* into a separate family Paeoniaceae and *Austrobaileya* into a separate family Austrobaileyaceae has been supported by studies of wood anatomy. The genus *Illicium* has been separated from Winteraceae because of unilacunar nodes, continuous pseudosiphonostele and the absence of granular material in stomatal depressions.

Trichomes

Trichomes have been of considerable help in Cruciferae (Schulz, 1936), especially in the genera *Arabis* and *Arabidopsis*. Trichome characters are very useful in the large genus *Astragalus* (with more than 2000 species). The Himalayan species *Hedera nepalensis* is distinguished from its European relative *H. helix* in having scaly trichomes as against stellate in the latter. In the family Combretaceae trichomes are of immense significance in classification of genera, species or even varieties (Stace, 1973). Trichomes are also diagnostic characters for many species of *Vernonia* (Faust and Jones, 1973).

Epidermal Features

Epidermal features are also of considerable taxonomic interest (SEM epidermal features are discussed under ultrastructure and micromorphology). Prat (1960) demonstrated that one can distinguish a **Festucoid type** (simple silica cells, no bicellular hairs) and **Panicoid type** (complicated silica cells, bicellular hairs) of epidermis in grasses. Stomatal types are distinctive of certain families such as Ranunculaceae (**anomocytic**), Brassicaceae (**anisocytic**), Caryophyllaceae (**diacytic**), Rubiaceae (**paracytic**), and Poaceae (**graminaceous**). Closely related families Acanthaceae and Scrophulariaceae are distinguished by the presence of diacytic stomata in the former as against anomocytic in the latter. The stomatal features, however, are not always reliable. In *Streptocarpus* (Sahasrabudhe and Stace, 1979) cotyledons have anomocytic while mature organs have anisocytic stomata. In *Lippia nodiflora* the same leaf may show anomocytic, anisocytic, diacytic and paracytic stomata (Pant, 1964).

Leaf Anatomy

The florets of Poaceae are reduced and do not offer much structural variability. Leaf anatomy has been of special taxonomic help in this family. The occurrence of the C-4 pathway and its association with **Kranz anatomy** (dense thick-walled chlorenchymatous bundle sheath, mesophyll simple), has resulted in revised classifications of several genera of grasses. Melville (1962, 1983) developed his gonophyll theory largely on the basis of study of venation pattern of leaves and floral parts. The rejection of *Sanmiguelia* and *Furcula* as angiosperm fossils from the Triassic has largely been on the basis of a detailed study of the venation pattern of leaves (Hickey and Doyle, 1977). The more recent rediscovery of *Sanmiguelia* from the Upper Triassic of Texas (Cornet 1986, 1989) points to presumed angiosperm incorporation of features of both monocots and dicots. Discovery of the Late Triassic *Pannaulika* (Cornet, 1993) from the Virginia-North Carolina border has reopened the possibilities of Triassic origin of angiosperms.

Floral Anatomy

Floral anatomy has been one of the thoroughly explored areas, with significant contributions to the phylogeny of angiosperms. Vascular traces in the carpels of various genera of the family Ranunculaceae have confirmed the origin of achene (*Ranunculus, Thalictrum*, etc.) from follicle (*Delphinium, Aquilegia*, etc.) through successive reduction in the number of ovules ultimately to one. The additional traces which would have gone to other ovules, now aborted, can be observed in many genera. There thus is no justification for Hutchinson's separation of achene-bearing genera and follicle-bearing genera into separate families Ranunculaceae and Helleboraceae respectively.

Melville (1962, 1983) developed his **gonophyll theory** after studying the vasculature of carpel and other floral parts through the clearing technique. He believed the angiosperm carpel to be a modified dichotomous fertile branch adnate to the petiole of a leaf. Sporne (1971) cautioned against such a drastic conclusion citing the example of bathroom loofah.

The genus *Melandrium* was segregated from *Silene* on the basis of the ovary being unilocular as against partly septate in *Silene*. Detailed floral anatomy revealed that in all the species of both genera the ovary is multilocular, at least in the early stages of development. The septa break down to various degrees in different species as the ovary develops. Thus structurally the ovaries are similar. The two genera were consequently merged into the single genus *Silene*.

The inferior ovary in angiosperms has been formed in two ways: **appendicular origin** (formed by fusion of calyx, corolla and their traces to ovary wall; in this case all vascular traces have normal orientation, i.e. phloem towards the outside) or by **axial invagination** (formed by depression of the thalamus; the inner vascular traces have reverse orientation, i.e.

phloem towards the inside). Studies on floral anatomy have confirmed that in a large majority of families the inferior ovary is of appendicular origin. Only in a few cases (*Rosa*, Cactaceae etc.) is the origin by axial invagination of the thalamus.

Floral anatomy has also supported the inclusion of *Acer negundo* under *Acer*, and does not support its separation into a distinct genus *Negundo*. Although this species is specialised in having a dioecious habit and anemophily, anatomy of the flower shows unspecialised features of other species. Floral anatomy also supports the separation of *Menyanthes* from Gentianaceae into a distinct family Menyanthaceae. The genus *Centella* is separated from *Hydrocotyle* on the basis of inflorescence being a cyme, and ovules receiving vascular supply from alternate bundles. In *Hydrocotyle* the inflorescence is an umbel and ovules receive vascular supply from fusion of two adjacent bundles. *Paeonia* is a classical example of a genus, which was removed from family Ranunculaceae into a distinct family Paeoniaceae. The separation has been supported by evidence from morphology, embryology and chromosomes. Floral anatomy also supports this separation, as both sepals and petals have many traces, carpels have five traces and the stamens are centrifugal.

III. EMBRYOLOGY

Embryology has made a relatively lesser contribution in understanding taxonomic affinities. This is primarily because of long preparatory work needed for embryological studies. More often the study of hundreds of preparations may reveal just a single embryological characteristic of any significance. It may take many years of laborious and painstaking research to study even a few representatives of a family. The embryological features of major significance include microsporogenesis, development and structure of ovule, embryo sac development, endosperm and embryo development.

Families Marked Out by Distinct Embryological Features

A number of families of angiosperms are characterised by a unique embrylogical features found in all members. These include:

1. **Podostemaceae**. The family which includes perennial aquatic herbs has a unique embryological feature in the formation of a **pseudoembryo sac** due to the disintegration of nucellar tissue. The family is also characterised by the occurrence of pollen grains in pairs, bitegmic tenuinucellate ovules, bisporic embryo sac, solanad type of embryogeny, prominent suspensor haustoria, and absence of triple fusion and consequently endosperm.

2. **Cyperaceae.** Family Cyperaceae is characterised by the formation of only one microspore per microspore mother cell. Following meiosis, of the four microspore nuclei formed, only one gives rise to pollen grain. Besides Cyperaceae only Epacridaceae in a few members shows the degeneration of three microspore nuclei. Cyperaceae is distinct from these taxa in pollen shedding at the 3-celled stage, as against the 2-celled stage shedding in Epacridaceae.

3. **Onagraceae.** The family is characterised by *Oenothera* type of embryo sac, not found in any other family except as an abnormality. This type of embryo sac is 4-nucleate and is derived from the micropylar megaspore of the tetrad formed.

Specific Examples of Role of Embryological Data

There are a few examples of embryological data having been very useful in the interpretation of taxonomic affinities:

1. **Trapa.** The genus *Trapa* was earlier (Bentham and Hooker, 1883) included under the family Onagraceae. It was subsequently removed to the family Trapaceae (Engler and Diels, 1936; Hutchinson, 1959, 1973) on the basis of distinct aquatic habit, two types of leaves, swollen petiole, semiepigynous disc and spiny fruit. The following embryological features support this separation: i) pyramidal pollen grains with 3 folded crests (bluntly triangular and basin shaped in Onagraceae); ii) ovary semi-inferior, bilocular with single ovule in each loculus (not inferior, trilocular, with many ovules); iii) *Polygonum* type of embryo sac (not *Oenothera* type); iv) endosperm absent (not present and nuclear); v) embryo solanad type (not onagrad type); vi) one cotyledon extremely reduced (the two not equal); and vii) fruit large one-seeded drupe (not loculicidal capsule).

2. **Paeonia.** The genus *Paeonia* was earlier included under the family Ranunculaceae (Bentham and Hooker; Engler and Prantl). Worsdell (1908) suggested its removal to a distinct family, Paeoniaceae. This was supported on the basis of centrifugal stamens (Corner, 1946), floral anatomy (Eames, 1961) and chromosomal information (Gregory, 1941). The genus as such has been placed in a distinct monogeneric family, Paeoniaceae, in all modern systems of classification. The separation is supported by the following embryological features: i) centrifugal stamens (not centripetal); ii) pollen with reticulately pitted exine with large generative cell (not granular, papillate and smooth, small generative cell); iii) unique embryogeny in which early divisions are free nuclear forming a coenocytic stage, later only the peripheral part becomes cellular (not onagrad or solanad type); and iv) seed arillate.

3. **Exocarpos.** The genus *Exocarpos* (sometimes misspelled *Exocarpus*) is traditionally placed under the family Santalaceae. Gagnepain and Boureau (1947) suggested its removal to a distinct family, Exocarpaceae,

near Taxaceae under Gymnosperms on the basis of articulate pedicel, 'naked ovule' and presence of a pollen chamber. Ram (1959) studied the embryology of this genus and concluded that the flower shows the usual angiospermous character, the anther has a distinct endothecium and glandular tapetum, pollen grains shed at the 2-celled stage, embryo sac of the *Polygonum* type, endosperm cellular, and the division of zygote transverse. This confirms that the genus *Exocarpos* is undoubtedly an angiosperm and a member of the family Santalaceae, with no justification for its removal to a distinct family. The genus is as such placed in Santalaceae in all major systems of classification.

4. **Loranthaceae**. The family Loranthaceae is traditionally divided into two subfamilies Loranthoideae and Viscoideae largely on the basis of presence of a calyculus below the perianth in the former and its absence in the latter. Maheshwari (1964) noted that the Loranthoideae has triradiate pollen grains, *Polygonum* type of embryo sac, early embryogeny is biseriate, embryo suspensor present, and viscid layer outside the vascular supply in fruit. As against this Viscoideae has spherical pollen grains, *Allium*-type of embryo sac, early embryogeny many-tiered, embryo suspensor absent, and viscid layer inside the vascular supply of fruit. He thus advocated separation of the two as distinct families Loranthaceae and Viscaceae. The separation was accepted by Takhtajan (1980, 1987, 1997), Dahlgren (1980), Cronquist (1981, 1988) and Thorne (1981, 1992).

Palynology

The pollen wall has been a subject of considerable attention, especially in an attempt to establish the evolutionary history of angiosperms. Some families, such as Asteraceae, show different types of pollen grains (**eurypalynous**), whereas several others have a single morphological pollen type (**stenopalynous**). Such stenopalynous groups are of considerable significance in systematic palynology.

The pollen grain wall is made of two principal layers, outer **exine** and inner **intine**. The former is further differentiated into two layers: outer **ektexine** or **sexine** and inner **endexine**. The sexine is further distinguished into basal **foot layer, baculum** and **tectum**. The exine may be inaperturate (without an aperture), monocolpate (single slit), tricolpate (three slits), or multiporate (many pores). Fossil studies over the last three decades have confirmed the monosulcate pollen of *Clavitopollenites* described (Couper, 1958) from Barremian and Aptian strata of the Early Cretaceous of southern England (132 to 112 Ma) to be the oldest recorded angiosperm fossil with distinct sculptured exine, resembling the pollen of extant genus *Ascarina*. Brenner and Bickoff (1992) recorded similar but inaperturate pollen grains from the Valanginian (ca 135 Ma) from the Helez formation of Israel, now considered to be the oldest record of angiosperm fossils (Taylor and Hickey,

1996). This last discovery has led to the belief that the earliest angiosperm pollen were without an opening, the monosulcate types developing later. Many claims of angiosperm records before the Cretaceous were made, but largely rejected. Erdtman (1948) described *Eucommiidites* as a tricolpate dicotyledonous pollen grain from the Jurrassic. This, however, had bilateral symmetry instead of the radial symmetry of angiosperms (Hughes, 1961) and a granular exine with gymnospermous laminated endexine (Doyle et al., 1975).

Among examples of the role of pollen grains is *Nelumbo* whose separation from Nymphaeaceae into a distinct family, Nelumbonaceae, is largely supported by the tricolpate pollen of *Nelumbo* as against the monosulcate condition in Nymphaeaceae.

Brenner (1996) proposed a new model for the evolutionary sequence of angiosperm pollen types. The earliest angiosperm pollen (from the Valanginian or earlier) was small, circular, tectate-columellate and without an aperture. In the Hauterivian possible thickening of the intine coupled with thickened endexine and evolution of the sulcus occurred. A considerable diversification of these monosulcate pollens occurred in the Barremian. Tricolptate pollen evolved in northern Gondwana in the lower Aptian.

IV. MICROMORPHOLOGY AND ULTRASTRUCTURE

Although widely used for lower plants, electron microscopy has been a comparatively new approach for flowering plants. The finer details of external features (**micromorphology**) have been explored in recent years by scanning electron microscopy (**SEM**), whereas the minute details of cell contents (**ultrastructure**) have been discerned through transmission electron microscopy (**TEM**). On average, the resolution power of SEM is 250A (20 times as good as optical microscope, but 20 times lesser than TEM). Behnke and Barthlott (1983) have made extensive studies of SEM and TEM characters. In most of the examples studied by electron microscopy (EM) the characters, proved to be stable and unaffected by environmental conditions.

Micromorphology

SEM studies have been made primarily on pollen grains, small seeds, trichomes and surface features of various organs. In most of these organs (except pollen grains) the studies involved the epidermis. The value of epidermal studies lies in the fact that an epidermis covers almost all organs and is always present, even in herbarium specimens. The epidermis is thick and stable in SEM preparations and is little affected by environment. However, it is important to note that only comparable epidermis should be studied (e.g. petals of all plants, leaves of all plants, not petals of some and leaves of others). The micromorphology of the epidermis includes the following aspects.

1. Primary sculpture   This refers to the arrangement and shape of cells. The **arrangement of cells** is specific for several taxa. In Papaveraceae seed coat cells by a particular arrangement form a reticulate supercellular pattern, which is a family character. There is specific distribution of long and short cells over the veins in the family Poaceae. The **shape of cells** is mainly determined by the **outline** of the cells, **boundaries** of the walls (which may be straight or undulated (S-, U-, omega-, V-types)), **relief** (channelled or raised boundaries), and cell **curvature** (flat, convex or concave).

2. Secondary sculpture   The secondary sculpture is formed by the fine relief of the outer wall of the **cuticle**, which may be striate, reticulate or micro-papillate (verrucose). All members of Urticales have curved trichomes with silicified cuticular striations at the base, and micro-papillations on the trichome body. This single character of trichomes allows for precise circumscription of the order Urticales (Barthlott, 1981). Loasiflorae is circumscribed by unicellular complexly hooked trichomes.

3. Tertiary sculpture   This is formed by secretions such as waxes and shows a variety of patterns. Secondary and tertiary sculpturing are mutually exclusive as the presence of waxes would invariably mask the cuticle; the cuticle would be visible only if there are no wax deposits. Winteraceae have a particular type and distribution of wax-like secretions (alveolar material not soluble in lipid solvents) on their stomata, similar to gymnosperms, and absent in all other angiosperms.

In monocots orientation and pattern of epicuticular waxes seem to provide a new taxonomic character of high systematic significance. Four types of wax patterns and crystalloids have been distinguished (Barthlott and Froelich, 1983):

(i)   **Smooth wax layers** in the form of thin films, common in angiosperms.

(ii)   **Non-oriented wax crystalloids** in the form of rodlets or platelets with no regular pattern. These are common in dicots and liliiflorous groups of monocots.

(iii)   **Strelitzia wax type** with massive compound wax projections composed of rodlet-like subunits that form massive compound plates around the stomata. This wax type is found in Zingiberiflorae, Commeliniflorae, and Areciflorae. It is also found in Velloziales, Bromeliales, and Typhales, which further differ from other Liliiflorae in a starchy endosperm.

(iv)   **Convallaria wax type** with small wax platelets arranged in parallel rows, which cross the stomata at a right angle and form a close circle around each polar end of the stoma—like lines of an electromagnetic field. This type is restricted to Liliiflorae only.

Cactaceae is a huge family commonly divided into three subfamilies, of which 90 per cent of the species are included in Cactoideae whose classification is difficult because of uniform floral characters, pollen morphology and plasticity. Barthlott and Voit (1979) analysed 1050 species and 230 genera by SEM for seed-coat structure in the family Cactaceae. The simple unspecialised testa of Pereskoideae supports its ancestral position. Opuntioideae has a unique seed with a hard aril, thus confirming its isolated position, also indicated by pollen morphology. Cactoideae shows complex diversity confirming its advanced position and subtribes have been recognised based on seed-coat structure, each subtribe having distinctive features. Thus the genus *Astrophytum* has been transferred from Notocactinae to Cactinae.

Orchidaceae is another large family with complicated phylogenetic affinities. Minute **'dust seeds'** show microstructural diversity of the seed-coat. Studies of over 1000 species (Barthlott, 1981) have helped in better subdivision into subfamilies and tribes. The Barthlott study also supports the merger of Cyperipediaceae with Orchidaceae.

Ultrastructure

Ultrastructure studies of angiosperms have provided valuable taxonomic information about the phloem tissue, mainly sieve tube elements. Besides this, information has also come from studies of seeds.

1. Sieve-element plastids Studies on sieve-element plastids were first initiated by Behnke (1965) in the family Dioscoreaceae. Since then nearly all angiosperm families have been investigated for the taxonomic significance of these plastids. All sieve-element plastids contain starch grains differing in number, size and shape. The protein accumulates in specific plastids in the form of crystalloids and filaments. Thus two types of plastids are distinguished: **P-type** which accumulate proteins and **S-type** which do not accumulate proteins. Starch accumulation is of no primary importance in classification, since it may be present or absent in both types of plastids. P-type plastids are further divided into six subtypes (Behnke and Barthlott, 1983) (Fig. 8.1):

   (i) **PI-subtype**. The plastids contain single crystalloids of different sizes and shapes and/or irregularly arranged filaments. This subtype is thought to be the most primitive in flowering plants, mainly Magnoliales, Laurales and Aristolochiales.

   (ii) **PII-subtype**. This subtype contains several cuneate crystalloids oriented towards the centre of the plastid. All monocots investigated contain this subtype. It is significant to note that only members of dicots with this subtype, *Asarum*, and *Saruma* of Aristolochiaceae are widely regarded among the most primitive members of dicots, a possible link between monocots and dicots.

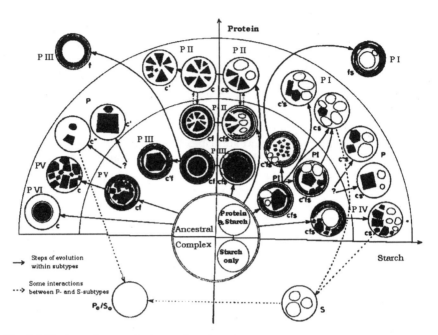

Fig. 8.1 Various forms of sieve-tube plastids and their possible evolution (after Behnke and Barthlott, 1983).

(iii) **PIII-subtype**. This subtype contains a ring-shaped bundle of filaments. PIII-subtype is confined to Centrospermae (Caryophyllales) and the removal of Bataceae and Gyrostemonaceae has been supported by the absence of this subtype in these families. Further, forms are recognised based on the presence or absence of crystalloids into **PIIIa** (globular crystalloid), **PIIIb** (hexagonal crystalloid) and **PIIIc** (without crystalloid). Based on the distribution of these forms Behnke (1976) proposed division of the order into three family groups which exactly correspond to the three suborders Caryophyllineae, Chenopodineae and Phytolaccineae earlier established by Friedrich (1956). Whereas Takhtajan had recognised these three suborders in his 1983 revision, in his final revision of his classification he merged Phytolaccineae with Caryophyllineae, thus recognising only two suborders, Caryophyllineae and Chenopodineae. Of the three orders recognised in Caryophyllideae of Takhtajan, only Caryophyllales contains PIII-subtype plastids while the other two orders, Polygonales and Plumbaginales, contain S-type plastids. Behnke (1977) as such advocated retention of only Caryophyllales under Caryophyllideae and removal of the other two orders to subclass Rosidae whose members also contain S-type plastids. The suggestion was not accepted by Takhtajan (1987) and Cronquist (1988) who retained all the three orders under

Caryophyllideae. Takhtajan, however, placed the three orders under separate superorders (Cronquist does not recognise superorders). On further intensive studies of plastids within the group, Behnke (1997) advocated removal of the genus *Sarcobatus* from the family Chenopodiaceae on the basis of presence of **PIIIcf** plastids and absence of **PIIIf**, which are characteristic of family Chenopodiaceae. He places the genus in an independent family, Sarcobataceae.

(iv) **PIV-subtype**. The plastid contains a few polygonal crystalloids of variable size. This subtype is restricted to the order Fabales.

(v) **PV-subtype**. The plastid contains many crystalloids of different sizes and shapes. This subtype is found in the order Ericales and family Rhizophoraceae.

(vi) **PVI-subtype**. The plastid contains a single circular crystalloid. This subtype is found in family Buxaceae.

2. Dilated Cisternae

Dilated Cisternae (DC) were first described by Bonnett and Newcomb (1965) as dilated sections of endoplasmic reticulum in root cells of *Raphanus sativus*. Originally found in Brassicaceae and Capparidaceae, DC have now been found in several other families of angiosperms but are concentrated in the order Capparales (Brassicaceae and Capparidaceae) and form a part of the character syndrome of this order. The DC may be utricular, irregular or vacuole-like in form with filamentous, tubular or granular contents. They have been proposed to be functionally associated with glucosinolates and myrosin cells found in this order.

3. Phloem (P-) proteins

P-proteins are found only in sieve elements of angiosperms and occur in the form of filaments or tubules. These assemble into large discrete bodies, and are not dissolved during maturation of the sieve element, unlike single membrane organelles. The composition and three-dimensional arrangement of these proteins exhibit taxonomic specificity. They are dispersed over the entire cell as the cell matures but in some dicots a single non-dispersive (crystalline) body of various shapes may be found in addition to dispersive ones. Crystalline bodies are absent in monocots. Their shape is often specific and thus of taxonomic importance. Globular crystalline bodies are found in Malvales and Urticales. Fabiflorae, which is characterised by PIV-subtype plastids, has spindle-shaped crystalline bodies in the family Fabaceae. The feature has supported the transfer of *Swartzia* to Fabaceae.

4. Nuclear inclusions

Nuclear inclusions in the form of protein crystals occur in phloem-and Ray parenchyma, mainly in the families of Asteridae. Five types of crystals have been differentiated. Structural differences have significance for classification within Scrophulariaceae and Lamiaceae (Speta, 1979). Protein crystals in sieve-tube elements have also been reported in Boraginaceae, and may prove useful.

5. Non-phloematic TEM characters   Protein bodies in seeds through TEM, SEM and dispersive X-ray techniques have demonstrated their significance if qualitative and quantitative aspects are both taken into account. Similarly, SEM studies of starch grains are also potential sources of information of taxonomic significance.

V. CHROMOSOMES

The chromosomes are the carriers of genetic information and as such have a considerable significance in evolutionary studies. Increased knowledge about chromosomes and their behaviour has largely been responsible for extensive biosystematic studies and development of the biological species concept. During the first quarter of the 20th century chromosomal data were relatively sparse. Such information has markedly increased over the last few decades, however, with ample useful information coming from studies of the banding pattern. Three types of chromosomal information have been of significance in taxonomy.

Chromosomal Number

Extensive records of chromosome numbers are available in the works of Darlington and Janaki-Ammal (1945), Darlington and Wylie (1955), Federov (1969) and Löve et al. (1977). International Association of Plant Taxonomy (IAPT) has also been publishing an **Index to Plant Chromosome Numbers** in its series *Regnum Vegetabile*. Between 1967 and 1977, the series published 9 volumes mostly forming annual lists of chromosome numbers. An updated server of Missouri Botanical Garden maintains the records of chromosome numbers and can be queried for online information about plant species. The chromosome counts are usually reported as diploid number $(2n)$ from mitosis of **sporophytic** tissue but when based on mitosis in **gametophytic** tissue or based on meiosis studies, counts are reported as haploid (n). The gametophytic chromosome number of diploid species is designated as **base-number** (x). In diploid species as such $n = x$, whereas in polyploid species n is in multiples of x. A hexaploid species with $2n = 42$ will thus have $n = 21$, $n = 3x$ and $2n = 6x$.

The chromosome number in angiosperms exhibits considerable variation. The lowest number $(n = 2)$ is recorded in *Haplopappus gracilis* (Asteraceae) and the highest $(n = 132)$ in *Poa littoroa* (Poaceae). The alga *Spirogyra cylindrica* also contains $n = 2$, whereas the record of the largest chromosome number $(n = 630)$ is found in *Ophioglossum reticulatum* (Pteridophytes). Such a range of variation, however, within nearly a quarter of a million species of angiosperms, may not be very significant in taxonomic delimitation, but there have been instances of the isolated role of studies on chromosomes. Raven (1975) provided a review of chromosome numbers at the family level in angiosperms. He concluded that the original base-number for angiosperms is $x = 7$ and that comparisons at the family level are valid only when the base-number (and not n or $2n$) is used. The family Ranunculaceae is dominated by genera with large chromosomes (and $x = 8$). The two genera *Thalictrum* and *Aquilegia* originally placed in two separate subfamilies or tribes (and even two separate families Ranunculaceae and Helleboraceae by Hutchinson, 1959, 1973 along with other achene-bearing and follicle-bearing genera respectively) are distinct in having small chromosomes (and $x = 7$) and as such have been segregated into a distinct tribe. The genus *Paeonia* with very large chromosomes (and $x = 5$) has been separated into a distantly related family Paeoniaceae, a placement which has been supported by morphological, anatomical and embryological data. Significant records in other families include Rosaceae with $x = 17$ in the subfamily Pomoideae, whereas other subfamilies have $x = 7, 8$ or 9. In Poaceae similarly the subfamily Bambusoideae has $x = 12$, whereas Pooideae has $x = 7$.

Spartina was for long placed in the tribe Chlorideae ($x = 10$) although its chromosomes ($x = 7$) were at variance. Merchant (1968) showed the genus to have in fact $x = 10$, thus securing its placement within Chloridae.

The classical study of the genus *Crepis* (Babcock, 1947) based separation from the closely related genera on chromosomal number and morphology. This led to the separation of the genus *Youngia* and merger of *Pterotheca* with *Crepis*. Similarly, in the genus *Mentha* which has small, structurally uniform chromosomes, the chromosome numbers provide strong support for subdivision into sections *Audibertia* ($x = 9$), *Pulegium* ($x = 10$), *Preslia* ($x = 18$) and *Mentha* ($x = 12$).

The duplication of chromosome numbers leading to **polyploidy** may prove to be of taxonomic significance. The grass genus *Vulpia* contains diploid ($2n = 14$), tetraploid ($2n = 28$) and hexaploid ($2n = 42$) species. The genus is divided into five sections, of which three contain only diploids, one diploids and tetraploids and one all three levels of ploidy. It is presumed that tetraploid and hexaploid species of *Vulpia* arose from diploid progenitors. The duplication of chromosome number of a diploid species may form a tetraploid (**autopolyploid**). Such a polyploid, however, does not show any or at most may show minor differences from the diploid species, and is rarely recognised as an independent taxonomic entity. The hybrid between two diploid species contains one genome from either parent and thus generally doesn't survive for failure to pair during meiosis. Hybridisation followed by duplication of chromosomes establishes a tetraploid (**allopolyploid; amphiploid**) with normal pairing as both genomes are in pairs. Such a tetraploid hybrid with distinct characteristics may be recognised as an independent species. A triploid hybrid between a diploid species and a tetraploid species may similarly not survive as a genome from a diploid parent would exhibit the problem of pairing at meiosis but the hybridisation followed by duplication leading to hexaploidy can form a perfectly normal independent species. Such facts have led to the detection of hybrids or confirmation of suspected hybrids. *Senecio* (Asteraceae) includes the diploid *S. squalidus* ($2n = 20$), the teraploid *S. vulgaris* ($2n = 40$) and the hexaploid *S. cambrensis* ($2n = 60$). The last is intermediate in morphology between the first two and is found in the area where these two grow. Additionally, sterile triploid hybrids between two species have been reported. It seems clear that *S. cambrensis* is an allohexaploid between the other two species (Stace, 1989). Similarly, based on chromosome number and karyotype, Owenby (1950) concluded that *Tragopogon mirus*, a tetraploid species, arose as an amphiploid between two diploid species, *T. dubius* and *T. porrifolius*.

Whereas a species generally shows a single chromosome number, certain populations or infraspecific taxa (subspecies, variety, forma) may sometimes show a different chromosome number (or even different chromosomal morphology). Such populations or infraspecific taxa constitute **cytotypes**.

Chromosomal Structure

Chromosomes show considerable variation in size, position of centromere and presence of secondary constriction. The chromosomes are commonly differentiated as **metacentric** (with centromere in middle), **submetacentric** (away from middle), **acrocentric** (near the end) or **telocentric** (at the end). The chromosomes are also characterised by their size. In addition the occurrence and position of **secondary constriction**, which demarcates a **satellite** is important in chromosomal identification and characterisation. The identification of satellites is often difficult, and especially when the secondary constriction is very long, a satellite may be counted as a distinct chromosome. This situation has often led to erroneous chromosome counts. The structure of the chromosome set (genome) in a species is termed the karyotype and is commonly diagrammatically represented in the form of an **ideogram** (Fig. 8.2) or **karyogram**. An analysis of a large number of studies

Fig. 8.2 Ideogram of the somatic complement of *Allium ampeloprasum*. Of the 32 somatic chromosomes 8 show secondary constriction (courtesy Prof. R.N. Gohil).

led to the conclusion that a **symmetrical karyotype** (chromosomes essentially similar and mainly metacentric) is primitive and an **asymmetric karyotype** (different types of chromosomes in a genome) advanced, the latter commonly found in plants with specialised morphological features such as *Delphinium* and *Aconitum*.

An interesting example of utilisation of chromosomal information is the family Agavaceae. The family contains about 18 genera such as *Agave* (and others formerly placed in Amaryllidaceae due to inferior ovary) and *Yucca* (and others formerly placed in Liliaceae due to superior ovary). These genera were shifted and brought into Agavaceae on the basis of great overall similarity. This was supported by the distinctive **bimodal karyotype** of Agavaceae consisting of 5 large chromosomes and 25 small ones. Rousi (1973) from his studies on the genus *Leontodon* showed that data on basic number, chromosome length, centromeric position and the occurrence of satellites provide evidence for the relegation of the former genus *Thrincia* ($x = 4$) as a section of subgenus *Apargia* along with section *Asterothrix* ($x = 4, 7$). The subgenus *Leontodon* is distinct with $x = 6$ or 7, and a different chromosome morphology.

Cyperaceae and Juncaceae were earlier placed far apart due to distinct floral structure. Both families have small chromosomes without distinctive centromeres, which may be diffuse or non-localised (**holocentric chromosomes**), and are now considered to be closely related. Such chromosomes do not depend on a discrete centromere for meiosis and mitosis and may undergo fragmentation with no deleterious effect. This may result in variable chromosomal counts. In the *Luzula spicata* group, chromosomal counts are reported to be $2n = 12, 14$, and 24. Interestingly the total chromosomal volume is the same and the higher chromosome number is the result of fragmentation (**agmatoploidy**) of these holocentric chromosomes. Different chromosome numbers may often occur in different cells of the same root-tip (**mixoploidy**). The occurrence of accessory chromosomes (known as **B-chromosomes**) in higher plants generally exerts no significant effect on morphology and thus is of little taxonomic importance. B-chromosomes in bryophytes, contrarily are very small (termed **m-chromosomes**) and often highly diagnostic.

In recent years a considerable breakthrough has been achieved in the study of **banding patterns** of chromosomes using Giesma and fluorochrome stains. Already different techniques such as **C-banding, G-banding, Q-banding** and **Hy-banding** are in use, and help in clearly distinguishing heterochromatic and euchromatic regions. C-banding is very useful in indicating the position of centromeres in cases where they cannot be identified by conventional staining. The technique of **silverstaining** has been developed to highlight **NOR** (nucleolar organising region). An interesting study of the chromosomes of top onion (variously recognised as *Allium cepa* var. *viviparum* or *A. fistulosum* var. *proliferum*) as also those of *A. cepa* and

A. fistulosum was done by Schubert, Ohle and Hanelt (1983). By Geisma banding pattern and silver-staining studies they concluded that some chromosomes of top onion resemble *A. cepa* and others resemble *A. fistulosum*. Of the two satellites one resembles either species. Top onion is as such a **pseudodiploid** with no homologous pair. The study confirmed that top onion is a hybrid between the two aforesaid parents, and thus would be better known as *A. × proliferum* (Moench) Schrad. (based on *Cepa proliferum* Moench), and not as a variety of either species. Interestingly the top onion owes its existence to the bulbils, which are produced in place of an inflorescence and ensure the multiplication of the hybrid, which is otherwise sterile.

Chromosomal Behaviour

The fertility of a plant is highly dependent on the ability of meiotic chromosomes to pair (**synapsis**) and their subsequent separation. Meiotic behaviour of chromosomes enables comparison between genomes to detect the degree of homology, especially when they are a result of hybridisation. Greater degree of genomic non-homology results in failure of pairing (**asynapsis**) or a loose pairing of chromosomes without chiasmata so that chromosomes fall apart before metaphase (**desynapsis**). In extreme cases the whole genome may fail to pair. The genome analysis of suspected hybrids has helped in establishing the parentage of several polyploid species. A diploid hybrid between two species generally exhibits failure of meiotic pairing due to non-homology of genomes resulting in hybrid sterility, but when hybridisation is followed by duplication of chromosomes to form a tetraploid hybrid, the latter shows normal pairing between the two genomes derived from the same parent and is generally fertile. A triploid hybrid may similarly be sterile but a hexaploid one fertile. Genome analysis has confirmed that the hexaploid *Senecio cambrensis* is allohexaploid between tetraploid *S. vulgaris* and diploid *S. squalidus*. Similarly, as already mentioned, the tetraploid *Tragopogon mirus* is the result of hybridisation between the two diploid species *T. dubius* and *T. porrifolius*. The most significant case, however, is the common bread wheat *Triticum aestivum*, a hexaploid with AABBDD genome. Genome analyses have confirmed that genome A is derived from the diploid *T. monococcum*, B from *Aegilops speltoides*, both genomes being represented in the tetraploid *T. dicoccum*. Genome D is derived from the diploid *Aegilops squarrosa*.

VII. CHEMOTAXONOMY

Chemotaxonomy of plants is an expanding field of study and seeks to utilise chemical information to improve classification of plants. Chemical evidence has, in fact, been used ever since man first began to classify plants as edible and inedible, obviously based on their chemical differences. Chemical information about medicinal plants in herbals published nearly four centuries back was concerned with localisation and application of physiologically active secondary metabolites such as saponins and alkaloids. Knowledge about chemistry of plants greatly increased during the 18th and 19th centuries. The greatest interest has been generated over the last 40 years, however, with the development of improved techniques for studying biological molecules, especially proteins and nucleic acids. In recent years interest has focused on the study of **allelochemy** and realisation of the concept that the animal kingdom and the plant kingdom have experienced a chemical **co-evolution**. Plants continuously evolve new defensive chemical mechanisms to save themselves from predators, and animals evolve methods to overcome these defences. In the process some plant species have developed animal hormones, thus disturbing the harmonal levels of animals if ingested.

A large variety of chemical compounds are found in plants and quite often the biosynthetic pathways producing these compounds differ in various plant groups. In many instances the biosynthetic pathways correspond well with existing schemes of classification based on morphology. In other cases the results are at variance, thus calling for revision of such schemes. The natural chemical constituents are conveniently divided as under:

1. *Micromolecules*—compounds with low molecular weight (less than 1000).
   (i) **Primary metabolites**—compounds involved in vital metabolic pathways: citric acid, aconitic acid, protein amino acids etc.
   (ii) **Secondary metabolites**—compounds which are the by-products of metabolism and often perform non-vital functions: non-proteinic amino acids, phenolic componds, alkaloids, glucosinolates, terpenes etc.
2. *Macromolecules*—compounds with high molecular weight (1000 or more).
   (i) **Non-semantide macromolecules**—compounds not involved in information transfer: starches, celluloses etc.
   (ii) **Semantides**—information carrying molecules: DNA, RNA and proteins.

Primary Metabolites

Primary metabolites include compounds, which are involved in vital metabolic pathways. Most of them are universal in plants and of little taxonomic importance. Aconitic acid and citric acid, first discovered from *Aconitum* and *Citrus* respectively, participate in Krebs cycle of respiration and are found in all aerobic organisms. The same is true of the 22 or so amino acids, and the sugar molecules, which are involved in the Kalvin cycle of photosynthesis. The quantitative variations of these primary metabolites may, however, be of taxonomic significance sometimes. In *Gilgiochloa indurata* (Poaceae) **alanine** is the main amino acid in leaf extracts, **proline** in seed extracts and **asparagine** in flower extracts. Rosaceae is similarly rich in arginine.

Secondary Metabolites

Secondary metabolites perform non-vital functions and are less widespread in plants compared to primary metabolites. These are generally the by-products of metabolism. They were earlier considered to be waste products, having no important role. Recently, however, it was realised that they are important in chemical defence against predators, pathogens, allelopathic agents and also help in pollination and dispersal (Swain, 1977). Gershenzon and Mabry (1983) have provided a comprehensive review of the significance of secondary metabolites in higher classification of angiosperms. The following major categories of secondary metabolites are of taxonomic significance:

1. Non-proteinic amino acids   A large number of amino acids not associated with proteins are known (more than 300 or so). Their distribution is not universal but specific to certain groups and thus hold promise for taxonomic significance. **Lathyrine** is thus known only from *Lathyrus*. **Canavanine** occurs only in Fabaceae and is shown (Bell, 1971) to be a protection against insect larvae. These amino acids are usually concentrated in storage roots and as such root extracts are generally used for their study.

2. Phenolics   Phenolic compounds form a loose class of compounds, based upon a phenol (C_6H_5OH). **Simple phenols** are made of a single ring and differ in position and number of OH groups. These are widely distributed in the plant kingdom; common examples are catechol, hydroquinone, phloroglucinol and pyrogallol. Coumarins, a group of natural phenolics, have a characteristic smell. The crushed leaves of *Anthoxanthum odoratum* can thus be identified by this characteristic smell.

Flavonoids, the more extensively studied compounds, are based on a flavonoid nucleus consisting of two benzene rings joined by a C_3 open or closed structure (Fig. 8.3). Common examples are flavonols, isoflavones, malvadins, anthocyanadins (often combining with different sugars and at different places to form various types of anthocyanins). **Anthocyanins** and **Anthoxanthins** are important pigments in the cell sap of petals providing red, blue (anthocyanins), and yellow (anthoxanthins) colours in a large number of families of angiosperms. These pigments are absent in some families and are replaced by highly different compounds, **betacyanins** and **betaxanthins** (together known as **betalains**), which consist of heterocyclic nitrogen-containing rings and have quite distinct metabolic pathways of synthesis. However, they carry the same functions as anthocyanins.

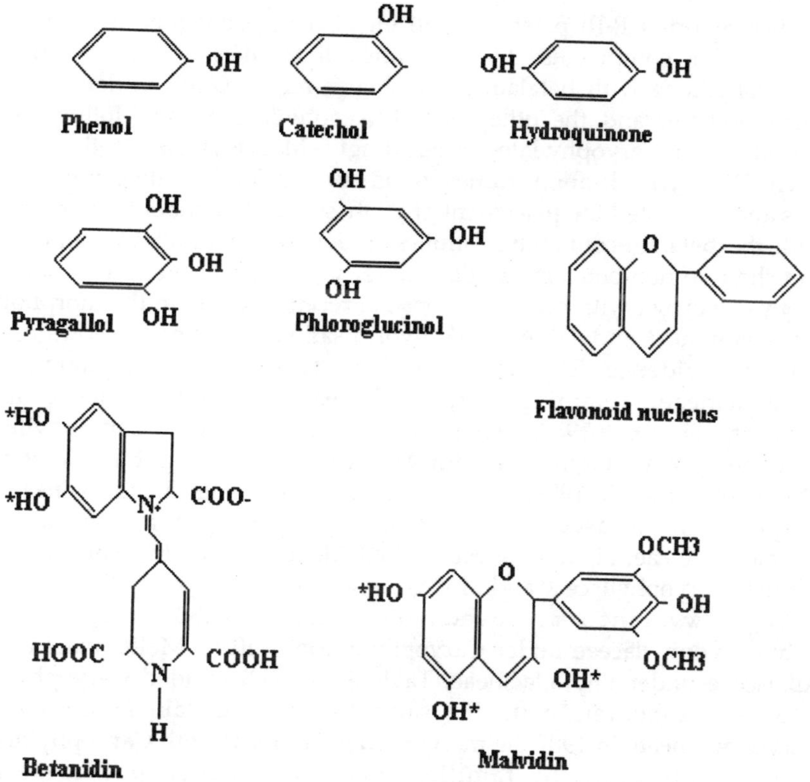

Fig. 8.3   Structure of important phenolic molecules.
       * indicates the position of sugar.

Betalains are mutually exclusive with anthocyanins, and concentrated in the traditional group Centrospermae of Engler and Prantl, now recognised as Caryophyllales. Of the nine families which contain betalains, seven were included in Centrospermae, Cactaceae placed in Cactales or Opuntiales and the ninth was placed in Sapindales. Traditional Centrospermae also included Gyrostemonaceae, Caryophyllaceae and Molluginaceae which lack betalains and contain anthocyanins instead. Mabry et al. (1963) on the basis of separate structure and metabolic pathways, suggested the placement of only betalain-containing families in Centrospermae, thus advocating the inclusion of Cactaceae and Didiereaceae and exclusion of Gyrostemonaceae, Caryophyllaceae and Molluginaceae. Whereas the inclusion of Cactaceae and Didieraceae was readily accepted (thus bringing all betalain-containing families into the same order, Centrospermae), exclusion of Caryophyllaceae and Molluginaceae was strongly opposed on the basis of structural data. This clash between orthodox and chemical taxonomy initiated renewed interest in the group. Behnke and Turner (1971) on the basis of ultrastructure studies reported P-III plastids in all members of Centrospermae and thus suggested a compromise by including all families within the subclass Caryophyllidae with betalain-containing families placed under the order Chenopodiales and the other two (Caryophyllaceae and Molluginaceae) placed under Caryophyllales. Interestingly, Mabry (1976) on the basis of DNA/RNA hybridisation studies found closer affinities between these families and suggested the placement of all these families under Caryophyllales with the betalain-containing families under the suborder Chenopodineae and the two non-betalain families under Caryophyllineae. This final compromise has met with mixed response in recent years with the morphological, anatomical and DNA/RNA hybridisation evidence overridding the betalain evidence. Thorne (1976) recognised three suborders (third Portulacineae), retaining Caryophyllaceae and Molluginaceae under Caryophyllineae. In 1983, he merged Chenopodineae with Caryophyllineae (also merging Molluginaceae with Aizoaceae, recognising it as a subfamily Molluginioideae). In 1992, he went back to establishing Chenopodineae and Caryophyllineae as distinct and also recognising a fourth suborder Achatocarpineae. He also re-established Molluginaceae to retain only two families Caryophyllaceae and Molluginaceae under Caryophyllineae. In 1997 (-http://www.inform.umd.edu/pb10/pb250/pb250.html) he retained only Caryophyllaceae under Caryophyllineae, shifting Molluginaceae near Aizoaceae under Phytollacineae. Takhtajan (1983) divided Caryophyllales into three suborders, with the same two non-betalain families under Caryophyllineae. In 1987, he merged Phytolaccineae with Caryophyllineae, thus including all the families except two (Chenopodiaceae and Amaranthaceae placed in Chenopodineae) in Caryophyllineae. In his 1997 classification, he abolished the suborders thus placing all the families together. Dahlgren (1983, 1989) and Cronquist (1981, 1988) never recognised suborders, the former placing the two—Caryophyllaceae and Molluginaceae —towards the beginning of the order Caryophyllales, whereas Cronquist placed them towards the end (see Table 8.1).

It is interesting to note that the betalains have also been reported in Basidiomycetes (Fungi), in some cases the same substance found in both fungi and angiosperms. The above studies on the significance of distribution of betalains in Centrospermae brings home the fact that chemical data are useful in taxonomic realignments when such accord with data from other fields. The significance is negligible when larger evidence from elsewhere contradicts the chemical evidence. Thus whereas no questions were ever asked about the removal of Gyrostemonaceae and the inclusion of Cactaceae and Didiereaceae, there has been no agreement about the removal of Caryophyllaceae and Molluginaceae as it goes against the evidence from morphology, anatomy, ultrastructure and DNA/RNA hybridisation. This also highlights the danger of relying too much on one type of evidence.

Studies on phenolic compounds have helped in solving some specific problems. Bate-Smith (1958) studied five phenolic characters of different

Table 8.1: Classification of order Centrospermae

Structural classification	*Chemical classification*	*Compromise classification*	*Thorne's classification (1997)*	*Dahlgren's classification (1989)*	*Cronquist's classification (1988)*	*Takhtajan's classification (1997)*
Centrospermae	*Chenopodiales*	*Caryophyllidae* / *Chenopodiales*	*Caryophyllanae* / *Caryophyllales*	*Caryophyllanae* / *Caryophyllales*	*Caryophyllidae* / *Caryophyllales*	*Caryophyllidae* / *Caryophyllanae* / *Caryophyllales*
Chenopodiaceae	Aizoaceae	Aizoaceae	*Achataocarpineae*	Molluginaceae	Phytolaccaceae	Phytolaccaceae
Amaranthaceae	Amaranthaceae	Amaranthaceae	Achatocarpaceae	Caryophyllaceae	Achatocarpaceae	Gisekiaceae
Nyctaginaceae	Basellaceae	Basellaceae	Portulacaceae	Phytolaccaceae	Nyctaginaceae	Agdestidacxeae
Phytolaccaceae	Cactaceae	Cactaceae	Hectorellaceae	Achatocarpaceae	Aizoaceae	Barbeuiaceae
Gyrostemonaceae	Chenopodiaceae	Chenopodiaceae	Basellaceae	Agdestidaceae	Didiereaceae	Achatocarpaceae
Aizoaceae	Didiereaceae	Didiereaceae	Didiereaceae	Basellaceae	Cactaceae	Petiveriaceae
Portulacaceae	Nyctaginaceae	Nyctaginaceae	Cactaceae	Portulacaceae	Chenopodiaceae	Nyctaginaceae
Basellaceae	Phytolaccaceae	Phytolaccaceae	*Phytolaccineae*	Stegnospermataceae	Amaranthaceae	Aizoaceae
Caryophyllaceae	Portulacaceae	Portulacaceae	Stegnospermataceae	Nyctaginaceae	Portulacaceae	Sesuviaceae
Molluginaceae			Phytolaccaceae	Aizoaceae	Basellaceae	Tetragoniaceae
			Nyctaginaceae	Halophytaceae	Molluginaceae	Stegnospermataceae
			Sarcobataceae	Cactaceae	Caryophyllaceae	Portulacaceae
			Aizoaceae	Didiereaceae		Hectorellaceae
			Halophytaceae	Hectorellaceae		Basellaceae
			Molluginaceae	Chenopodiaceae		Halophytaceae
			Chenopodiineae	Amaranthaceae		Cactaceae
			Chenopodiaceae			Didiereaceae
			Amaranthaceae			Molluginaceae
						Caryophyllaceae
						Amaranthaceae
						Chenopodiaceae
Cactales						*Gyrostemonanae* / *Gyrostemonales*
Cactaceae						Gyrostemonaceae
Sapindales	*Caryophyllales*	*Caryophyllales*	*Caryophyllineae*			
Didiereaceae	Caryophyllaceae	Caryophyllaceae	Caryophyllaceae			
	Molluginaceae	Molluginaceae				

sections in the genus *Iris*. The chemical evidence supported the division into various sections, but *I. flavissima*, originally placed in the section *Pogoniris*, resembled species of the section *Regelia* on the basis of phenolic characteristics. Chromosomal evidence also supported this transfer.

The technique of two-directional paper chromatography, which brings about a more pronounced separation of flavonoids, has proved very useful in taxonomic studies. *Hymenophyton* (Bryophytes) was considered by some researchers to be a monotypic genus, but by others to include two species. Markham et al. (1976) on the basis of rapid flavonoid extraction, two-dimensional chromatographic analysis and identification (Fig. 8.4) concluded that the genus contains two distinct species, *H. leptodotum* and *H. flabellatum*, and there is no justification for their merger.

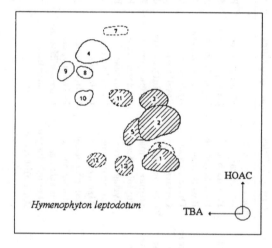

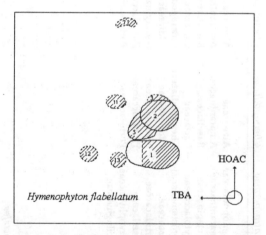

Fig. 8.4 Two-dimensional paper chromatograms of the flavonoids in two species of *Hymenophyton* (after Markham et. al., 1976).

Similar studies in the genus *Baptisia* (Fabaceae) by Alston and Turner (1963) have been very useful in the detection of hybridisation. Each species of the genus has a distinctive spectrum of flavonoids, and the hybrid can easily be identified by the combination of flavonoid pattern of both parental species in the suspected hybrid. It is interesting to note that the ten taxa recognised (four parental and six hybrid), could not be differentiated on the basis of morphological or biochemical characters alone, but a combination of both enabled a complete separation. The flavonoids in these studies were extracted from either flower or leaf.

3. Alkaloids Alkaloids are organic nitrogen-containing bases, usually with a heterocyclic ring of some kind. **Nicotine** (*Nicotiana*), **lupinine** (*Lupinus*), **quinine** (*Cinchona*) and **ephedrine** (*Ephedra*) are the familiar examples. Their distribution is often specific and thus taxonomically significant. **Morphine** is present only in opium poppy (*Papaver somniferum*). Mears and Mabry (1971) in studies on the family Fabaceae observed that the alkaloid **hystrine** occurs only in three genera *Genista*, *Adenocarpus* (both belonging to Genistae) and *Ammodendron* (originally placed in Sophorae). The latter, however, lacks **matrine**, characteristic of Sophorae. This indicates that the transfer of the last genus also to Genistae is warranted. Families Papaveraceae and Fumariaceae are closely related. This affinity is supported by the occurrence of the alkaloid **protopine** in both.

Gershenzon and Mabry (1983) reported that tropane alkaloids of Solanaceae and Convolvulaceae are similar suggesting a close relationship. The families are placed in the same order in recent systems. Papaveraceae, earlier grouped with Cruciferae and Capparaceae, is now removed to nearer Ranunculales on the basis of absence of glucosinolates and presence of benzylisoquinolene. Nymphaeaceae and Nelumbonaceae differ in that the former lacks benzylisoquinolene alkaloids.

4. Glucosinolates Glucosinolates are mustard oil glucosides found in the order Capparales. Originally Cruciferae, Capparaceae, Papaveraceae and Fumariaceae were placed in the same order, Rhoeadales. Chemical and other evidence, however, supported the placement of Cruciferae and Capparaceae in the order Capparales (on the basis of presence of glucosinolates) and Papaveraceae and Fumariaceae in the order Papaverales (on the basis of absence of glucosinolates and presence of the alkaloid benzylisoquinolene). Bataceae and Gyrostemonaceae were once placed in Centrospermae (Caryophyllales) but subsequently removed due to the absence of betalains. This removal was supported by the presence of glucosinolates, which are absent in Caryophyllales.

5. Terpenes Terpenes include a large group of compounds derived from the mevalonic acid precursor and are mostly polymerized isoprene derivatives. Common examples are **camphor** (*Cinnamomum*), **menthol** (*Mentha*), and **carotenoids**. They seem to have a definite role in the allelopathic effects of plants.

Terpenoids, the common group of terpenes, have been largely used in distinguishing specific and subspecific entities, geographic races and detection of hybrids. This has largely been possible by the technique of gas chromatography enabling qualitative as well as quantitative measure of chemical differences. Studies in *Citrus* have focused on determination of the origin of certain cultivars. Studies on *Juniperus viginiana* and *J. ashei* have refuted previous hypotheses about extensive hybridisation and introgression between the two species. Their distribution in *Pinus* has been used (Mirov, 1961) to understand relationships. *P. jeffreyi* has been considered a variety of *P. ponderosa*, but terpentine distribution showed that it strongly resembles the group Macrocarpae and not Australes to which *P. ponderosa* belongs. A major contribution of terpenoid chemistry has been the use of **sesquiterpene lactones** in the family Compositae. Many tribes within the family are characterised by distinct types of sesquiterpene lactones they produce. This helped to establish that genus *Vernonia* has two centres of distribution, one in the Neotropics and the other in Africa. Similarly, studies on *Xanthium strumarium* (McMillan et al., 1976) have thrown some light on the origin of Old World and New World populations. Old World populations produce xanthinin/ or xanthinosin, whereas the New World populations contain xanthinin or its stereoisomer xanthumin. Plants of the *chinense* complex (from Louisiana) contain xanthumin and are believed to be the source of introduced *chinense* populations in India and Australia.

Iridoids constitute another important group of terpenes (mostly monoterpene lactones). They are present in over 50 families and their presence is correlated with sympetaly, unitegmic tenuinucellate ovules, cellular endosperm and endosperm haustoria. Assuming that 'independent origin of several groups with this combination of independent attributes is unlikely', Dahlgren brought together all iridoid producing families. The occurrence of iridoids in several unrelated families, e.g. Hamamelidaceae and Meliaceae, however, suggests that iridoids could have arisen independently several times in the evolution of angiosperms. The occurrence of a distinctive iridoid **aucubin** in *Budleia* has been taken to support its transfer from Loganiaceae to Budleiaceae.

Cronquist (1977) proposed that chemical repellents had an important role in the evolution of major groups of dicots. The alkaloid Isoquinolene of Magnoliidae gave way to tannins of Hamamelidae, Rosidae and Dilleniidae, which in turn gave way to the most effective iridoids in Asteridae, the family Compositae developing the most effective sesquiterpene lactones.

Non-semantide macromolecules

These include complex polysachharides such as starches and celluloses. Starches are commonly found in the form of grains which may be concentric (*Triticum, Zea*) or eccentric (*Solanum tuberosum*) and present anatomical characteristics which can be seen under a microscope. Detailed studies of starch grains under SEM also hold promise for taxonomic significance.

Semantides

Semantides constitute macromolecules, which are primary constituents of living organisms and are involved in information transfer. Based on their position in the information transfer DNA is a **primary semantide**, RNA a **secondary semantide** and proteins the **tertiary semantides**. Semantides are popular sources of taxonomic information, and most of this information has come from proteins. Because of their complex structure, special techniques are necessary for the isolation, study and comparison of semantides. These methods include **serology, electrophoresis, amino acid sequencing** and **DNA hybridisation** as well as **DNA/RNA hybridisation**.

1. Serology The field of **systematic serology** or **serotaxonomy** had its origin towards the turn of the 20th century with the discovery of serological reactions and development of the discipline of immunology. Precipitin reactions were first reported by Kraus (1897). The technique was first applied by J. Bordet (1899) in his work on birds, when he reported that immune reactions are relatively specific and the degree of cross reactivity was essentially proportional to the degree of relationship among organisms. The present technique of serology is based on immunological reactions shown by mammals when invaded by foreign proteins. In the study of estimating relationships between plants, the plant extract of species A containing proteins (**antigens**) is injected into a mammal (usually a rabbit). The latter will develop **antibodies**, each specific to an antigen with which it forms a precipitin reaction, coagulating and thus making it non-functional. These antibodies are extracted from the body of the animal as **antiserum**. This antiserum is capable of coagulating all proteins in species A, but when mixed with the protein extract of species B, the degree of precipitin reaction would depend on the similarity between the proteins of the two species.

Antigens are mostly extracted from seeds and pollen. In early works crude total comparison of precipitin reactions was done but now more refined methods have been developed which can bring about individual antigen-antibody reactions. Major methods include:

(i) **Double-diffusion serology.** In this technique the antigen mixture and antiserum are allowed to diffuse towards one another in a gel (Fig. 8.5). The different proteins travel at different rates and thus the reactions occur at different places on the gel. This method allows comparison of precipitin reactions of several antigen mixtures from different taxa simultaneously on the same gel. In a modification of this method, antiserum is placed in a circular well surrounded by a ring of several wells containing the samples of antigens.

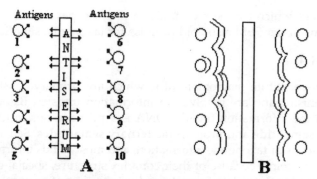

Fig. 8.5 Double-diffusion serology. A—Antigens and antibodies moving towards each other on the gel. 1-10 refer to antigen mixtures from ten different taxa. B—resultant precipitant lines.

(ii) **Immunoelectrophoresis.** In this technique the antigens are first separated unidirectionally in a gel by electrophoresis and then allowed to travel towards the antiserum (Fig. 8.6). This method enables a better separation of constituent reactions but has the limitation that only one antigen mixture can be handled on a single gel.

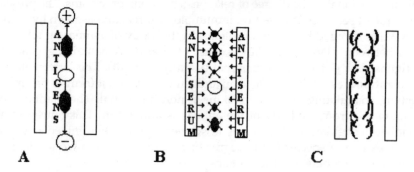

Fig. 8.6 Immunoelectrophoresis. A—Antigen separation by electrophoresis. B—Antibodies and separated antigens diffusing towards each other. C—Resultant precipitant lines.

(iii) **Absorption.** Protein mixtures from different species often contain a large number of common proteins, especially those involved in common metabolic processes. The antibodies for these common proteins (antigens) are first removed from the antiserum so that there is a more logical comparison of precipitin reactions.

(iv) **Radio-immunoassay (RIA).** In this technique the antibodies or antigens are labelled with radioactive molecules enabling their detection even when present in minute quantities.

(v) **Enzyme-linked immunosorbent assay (ELISA).** In this technique either the antibodies or antigens are labelled linked with enzymes, thus enabling detection even in very small quantities.

It must be noted that there are specific sites on proteins (**determinants**), which are capable of initiating production of immunoglobulins in specific cells of mammals. Determinants are regions consisting of 10-20 amino acids and one protein may comprise several different determinants and thus several **antigens**.

Extensive studies of the immunoelectrophoretic patterns of the genus *Bromus* were done by Smith (1972, 1983). Results showed that North American diploids of the genus are reasonably diverse. The study also highlighted that antisera raised from different species could provide different results. On the basis of serological studies Smith established the distinct identity of *B. pseudosecalinus*, previously recognised as a variety of *B. secalinus*. This separation was supported by cytological evidence also.

Serology may be done through comparison of protein mixtures or the comparison of single isolated and purified proteins. Schneider and Liedgens (1981) developed a complex but excellent procedure of monoclonal culture of antibodies, but unfortunately used this for construction of a '**phylogenetic tree**' not parallel with accepted evolutionary schemes. Fairbrothers (1983) cautioned that an evolutionary tree should not be constructed on the reactions of a single enzyme or a single species. Lee (1981) using purified protein for antigen and using different techniques concluded that *Franseria* (Asteraceae) should be merged with *Ambrosia*.

2. Electrophoresis Serology serves to compare the degree of similarity between the protein mixtures of different species and does not involve the identification of proteins. The separation and identification of proteins can be done by electrophoresis. Separation is based on the amphoteric properties of proteins whereby they are positively or negatively charged to various extents according to the pH of the medium, and will travel through gel at various speeds across a voltage gradient, usually carried out in a starch gel or acrylamide gel. The procedure involves homogenising the tissues (containing proteins) in a buffer solution. Pieces of filter paper are saturated with the homogenised extract by soaking and are then placed in a slit cut at about the centre of the gel. The current is run for a specific time, and the proteins run up to different points on the gel. The gel, usually 1 cm thick, is cut into three thin slices, each about 3 mm thick. These slices are subjected to different staining techniques and proteins are identified using various criteria. In **disc-electrophoresis** a gel of larger pores is placed over a gel of smaller pores. The former is used for crude separation and the latter for a complete separation.

In the technique of **isoelectric focusing** a gel of a single pore size is set up with a pH gradient (usually 3-10), so that proteins come to lie on the gradient corresponding to their iso-electric point. These can be subsequently separated more completely by disc-electrophoresis. Isoelectric focusing of **Rubisco** (Ribulose 1,5 diphosphate carboxylase) has been very useful in determining relationship between species of *Avena, Brassica, Triticum* and several other genera. It is an excellent protein for helping to evaluate hybridisation.

Electrophoretic studies have supported the origin of hexaploid wheat (*Triticum aestivum*). Johnson (1972) working on storage proteins showed that *T. aestivum* (AABBDD) and *T. dicoccum* (AABB) possess all proteins of the A genome of the diploid *T. monococcum* (AA). They also share proteins of the B genome of uncertain origin. The D genome is believed to have come from *Aegilops squarrosa* as evidenced by morphological and cytological data. By mixing proteins of *A. squarrosa* and *T. dicoccum* it was seen that the electrophoretic properties of the mixture closely resemble those of *T. aestivum*, thus proving the origin of the latter from the two previous species. Electrophoretic studies have also helped to assess species relationships in *Chenopodium* (Crawford and Julian, 1976), by combining data from flavonoids with proteins. A flavonoid survey of seven species showed that in some taxa the flavonoid data were fully compatible with interspecific protein differences, but in some cases did not agree. Thus *Chenopodium atrovirens* and *C. leptophyllum* had identical flavonoid patterns but could be distinguished by their different seed protein spectra. *C. desiccatum* and *C. atrovirens*, on the other hand, were closely similar in seed proteins but differed in flavonoids. Both flavonoid and protein evidence, however, distinguished *C. hians* from *C. leptophyllum*, thus providing support to their recognition as separate species. Vaughan et al. (1966) through the study of serology and electrophoresis have shown that *Brassica campestris* and *B. oleracea* are closer to each other than to *B. nigra*.

Electrophoresis has also made possible the separation of **allozymes** (different forms of the same enzyme with different alleles at one locus) and **isozymes** (or **isoenzymes** with different alleles at more than one locus). Barber (1970) showed that certain polyploids possess isozymes of all their progenitors plus some new ones. Backman (1964) crossed two strains of maize, each with three different isozymes. F1 possessed all six isozymes. The hybrids thus show molecular complementation. Studies of the genus *Tragopogon* have confirmed that the tetraploid *T. mirus* is a hybrid between two diploid species, *T. dubius* and *T. porrifolius*. Whereas the parental diploids were found to be divergent at close to 40 per cent of the 20 enzyme loci examined, the tetraploid hybrid possessed completely additive enzyme patterns. The evidence thus supported the recognition of a hybrid on the basis of morphological and chromosomal evidence.

3. Amino acid sequencing Since only 22 amino acids are known to be the constituents of proteins, the primary differences between the proteins result from different sequencing of amino acids in the polypeptide chain. It is now possible to break off the amino acids from the polypeptide chain one by one, identify each chromatographically and build up the sequence of amino acids step by step. **Cytochrome** *c* is the most commonly used molecule and out of 113 amino acids 79 vary from species to species, but alteration of even one of the other 34 destroys the functioning of the molecule. Being present in all aerobobic organisms, it is ideal for comparative studies. Boulter (1974) constructed a **cladogram** (Fig. 8.7) of 25 species of spermatophytes using the 'ancestral sequence method'. *Ginkgo biloba*, the only gymnosperm used, showed as isolated position in the cladogram. *Ginkgo* with an isolated phylogenetic position is no new discovery, but rather a long established fact. But the fact that amino acid sequencing also produces a similar cladogram establishes the significance of such studies in understanding phylogeny.

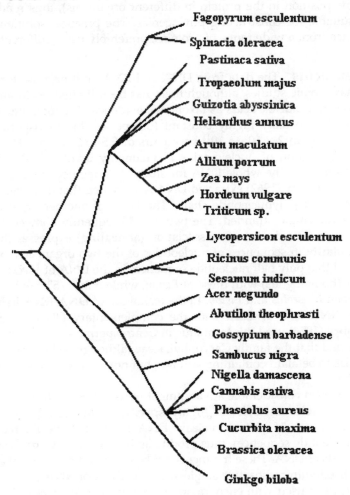

Fig. 8.7 Cladogram of 25 species of seed plants based on 'ancestral sequence method' used by Boulter (1974) (after Boulter).

Recent data from various fields have pointed to the merger of *Aegilops* with *Triticum*. Autran et al. (1979) on the basis of N-terminal amino acid sequencing supported this merger. In general, the number of amino acid differences is roughly parallel to the distance between the organisms in traditional classifications, suggesting that the method is broadly reliable. There are, however, certain contradictions. The number of differences between the cytochrome *c* of *Zea mays* and *Triticum aestivum* (both members of the same family Poaceae) is greater than between *Zea mays* and certain dicotyledons.

It has been found that cytochrome c and plastocyanin (another protein commonly used in amino acid sequencing studies) can exhibit a large number of **parallel substitutions** (identical changes from one amino acid to another at the same position in the protein in different organisms), thus rendering them unsuitable for constructing phylogenies. The practical solution is to use evidence from a wide range of proteins, preferably using different techniques.

4. Nucleic acids   The data from DNA and RNA hold promise for utilisation in taxonomic studies, although there has been limited exploration of information from nucleic acids as taxonomic evidence. The total amount of nuclear DNA is usually highly constant for species. In angiosperms the amount varies from 2×10^8 to 10^{11} base pairs or 0.5 to 254.8 pg. The early studies on utilisation of nucleic acids in taxonomy involved **DNA/DNA hybridisation** using the whole DNA for study. In a method developed by Bolton and Mecarthy (1962) extracted DNA is so treated that it converts into a single stranded polynucleotide chain. The DNA of another organism is similarly made single stranded. The two are subsequently allowed to hybridise in vitro. The degree of reassociation (**annealing**) expresses the degree of similarity in sequences of nucleotides of the two organisms. Bolton (1966) found that only half nucleotide sequences in the DNA of *Viccia villosa* are similar (homologous) with those of *Pisum*, while only 1/5th are homologous between *Phaseolus* and *Pisum*. In the technique of **DNA/RNA hybridisation** the RNA is hybridised with the complementary DNA of related plants. Mabry (1976) used this technique in Centrospermae (Caryophyllales) and concluded that the family Caryophyllaceae (although lacking betalains) is quite close to betalain-containing families, but not as close as the latter are to each other.

There has been considerable advancement in recent years. It is now possible to break (**cleave**) DNA at highly specific points using restriction endonucleases, each of which can cleave DNA at a highly specific recognition site (nucleotide sequence), thus producing highly characteristic **restriction fragments** of DNA. These fragments can be separated by electrophoresis and hybridised with radioactive single-stranded DNA or RNA probes. The method has been used with encouraging results in *Atriplex*, *Secale* and several other genera. Belford and Thomson (1979) using side-copy sequence hybridisation in *Atriplex* concluded that division into two subgenera in this genus is not correct.

Studies of DNA have largely been undertaken from chloroplast compared to the other two cellular genomes: nuclear and mitochondrial genomes. This is because chloroplast DNA (cpDNA) can be easily isolated and analysed. It is also not altered by evolutionary processes such as gene duplication and concerted evolution. It also has an added advantage in that it is highly conserved in organisation, size and primary sequence. Studies of cpDNA in *Brassica* have been helpful in establishing maternal progenitors of allopolyploids. It is significant to note that whereas the nuclear genome is inherited biparently, the chloroplast genome is inherited maternally. Thus the hybrid plant will possess the nuclear complement of both parents but only the cpDNA of the maternal plant.

9

Phenetic and Phylogenetic Methods

Systematics aims at developing classifications based on different criteria and often a distinct methodology is employed for the analysis of data. Data handling for establishing relationships between organisms often makes use of one of the two methods—**phenetic methods** and **phylogenetic methods**—often providing different types of classification. Distinction is sometimes also made between phylogenetic and evolutionary classification schemes.

I. PHENETIC METHODS: NUMERICAL TAXONOMY

Numerical taxonomy is a developing branch of taxonomy, which received a great impetus with the development and advancement of computers. This field of study is also known as **mathematical taxonomy** (Jardine and Sibson, 1971), **taxometrics** (Mayr, 1966), **taximetrics** (Rogers, 1963), **multivariate morphometrics** (Blackith and Reyment, 1971) and **phenetics**. The modern methods of numerical taxonomy had their beginning from the contributions of Sneath (1957), Michener and Sokal (1957), and Sokal and Michener (1958) which culminated in the publication of *Principles of Numerical Taxonomy* (Sokal and Sneath, 1963), with an expanded and updated version *Numerical Taxonomy* (Sneath and Sokal, 1973).

The last few decades have seen a forceful debate on the suitability of the **empirical approach** or **operational approach** in systematic studies. Empirical taxonomy forms classification on the basis of taxonomic judgement based on observational data and not assumptions. Operational taxonomy on the other hand is based on operational methods, experimentation to evaluate the observed data, before a final classification. Numerical taxonomy finds a balance between the two as it is both empirical and operational (Fig. 9.1).

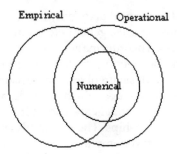

Fig. 9.1 Relationship between empirical, operational and numerical taxonomy (after Sneath and Sokal, 1973).

It must be remembered that numerical taxonomy does not produce new data or a new system of classification, but is rather a new method of organising data that could help in better understanding of relationships. Special classifications are based on one or a few characters or on one set of data. Numerical taxonomy seeks to base classifications on a greater number of characters from many sets of data in an effort to produce an entirely phenetic classification of maximum predictivity.

Principles of Numerical Taxonomy

The philosophy of modern methods of numerical taxonomy is based on ideas that were first proposed by the French naturalist Michel Adanson (1763). He rejected the idea of giving more importance to certain characters, and believed that natural taxa are based on the concept of similarity, which is measured by taking all characters into consideration. The principles of modern numerical taxonomy developed by Sneath and Sokal (1973) are based on the modern interpretation of the Adansonian principles and as such are termed **neo-Adansonian principles**. These **principles** of numerical taxonomy are enumerated below.

1. The greater the content of information in the taxa of a classification and the more characters it is based upon, the better a given classification will be.
2. A priori, every character is of equal weight in creating natural taxa.
3. Overall similarity between any two entities is a function of their individual similarities in each of the many characters in which they are being compared.
4. Distinct taxa can be recognised because correlations of characters differ in the groups of organisms under study.
5. Phylogenetic inferences can be made from the taxonomic structures of a group and from character correlations, given certain assumptions about evolutionary pathways and mechanisms.
6. Taxonomy is viewed and practised as an empirical science.
7. Classifications are based on phenetic similarity.

The methodology of numerical taxonomy involves the selection of operational units (populations, species, genera etc. from which the information is collected) and characters. The information from these is recorded and similarity (and/or distance) between units determined using various statistical formulae. The ultimate analysis involves comparison of similarity data and constructing diagrams or models, which provide a summary of the data analysis. These diagrams or models are used for final synthesis and better understanding of the relationships. The **major advantages** of numerical taxonomy over conventional taxonomy include:

1. Numerical taxonomy has the power to integrate data from a variety of sources, such as morphology, physiology, phytochemistry, embryology, anatomy, palynology, chromosomes, ultrastructure and micromorphology. This is very difficult to do by conventional taxonomy.

2. Considerable automation of the data processing promotes efficiency and the work can be handled by even less skilled workers.

3. Data coded in numerical form can be integrated with existing data-processing systems in various institutions and used for creation of descriptions, keys, catalogues, maps and other documents.

4. The methods, being quantitative, provide greater discrimination along the spectrum of taxonomic differences, and can provide better classifications and keys.

5. The creation of explicit data tables for numerical taxonomy necessitates the use of more and better described characters, which will necessarily improve conventional taxonomy as well.

6. The application of numerical taxonomy has posed some fresh questions concerning classification and initiated efforts for re-examination of classification systems.

7. A number of biological and evolutionary concepts have been reinterpreted, thus introducing renewed interest in biological research.

Numerical taxonomy aims at determining **phenetic relationships** between organisms or taxa. Cain and Harrison (1960) defined phenetic relationship as an *arrangement by overall similarity, based on all available characters without any weighting*. Sneath and Sokal (1973) define phenetic relationship as *similarity (resemblance) based on a set of phenotypic characteristics of the objects or organisms under study*. It is distinct from a cladistic relationship, which is an expression of the recency of common ancestory and is represented by a branching network of ancestor-descendant relationships. Whereas the phenetic relationship is represented by a **phenogram**, the cladistic relationship is depicted through a **cladogram**.

Operational Taxonomic Units

The first step in numerical analysis involves the selection of Operational Taxonomic Units (OTUs), the sample from which the data is collected.

Although it would be ideal to select individuals of a population as OTUs, practical considerations make it necessary to select the members of the next lower rank as OTUs. Thus OTUs for the analysis of a species would be various populations, for the study of a genus they would be different species, and for a family they would be different genera. It is not advisable, however, to use genera and higher ranks as OTUs as the majority of characters would show variation from one species to another and thus not be suitable for comparison. The practical solution would be to use one representative of each taxon. Thus if a family is to be analysed and its genera to be compared, the data from one representative species of each genus can be used for analysis. Once the OTUs are selected, a list of such OTUs is prepared.

Taxonomic Characters

A conventional definition of a taxonomic character is *a characteristic that distinguishes one taxon from another*. Thus white flowers may distinguish one species from another with red flowers. Thus white flower is one character and red flower another. A more practical definition espoused by numerical taxonomists defines character (Michener and Sokal, 1957) as *a feature which varies from one organism to another*. Thus by this second definition flower colour (and not white flower or red flower) is a **character**, and the white flower and red flower are its two **character states**. Some authors (Colless, 1967) use the term **attribute** for character state but the two are not always synonymous. When selecting a character for numerical analysis, it is important to select a **unit character**, which may be defined as *a taxonomic character of two or more states, which within the study at hand cannot be subdivided logically, except for the subdivision brought about by the method of coding*. Thus trichome type may be glandular or eglandular. A glandular trichome may be sessile or stalked. An eglandular trichome may similarly be unbranched or branched. In such a case a glandular trichome may be recognised as a unit character and an eglandular trichome as another unit character. On the other hand, if all glandular trichomes in OTUs are of the same type and all eglandular trichomes are of the same type, the trichome type may be selected as a unit character.

The first step in the handling of characters is to make a list of unit characters. The list should include all such characters concerning which information is desired. *A priori* all characters should be weighted equally (no weighting to be given to characters). Although some authors advocate that some characters should subsequently be assigned greater weightage than others (*a posteriori weighting*), such considerations generally get nullified when a large number of characters is used. It is generally opined that numerical studies should involve not less than 60 characters but more than 80 are desirable. For practical consideration there may be some characters concerning which information is not available (a large number of plants in a population are not in fruit) or the information is irrelevant (trichome type if a large number of plants are without trichomes), or the characters which show much variation within the same OTU (e.g. flower colour in *Lantana*). Such characters are omitted from the list. This constitutes ***residual weighting*** of characters. It is also important that only **homologous** characters be compared. The 'petals' of *Anemone* are modified sepals and thus not homologous with the petals of *Ranunculus* and hence not comparable.

Coding of Characters

The huge amount of data generally handled in numerical studies makes use of computers essential. Before the data is fed into the computer it should be suitably coded. The characters most suitable for computer handling are **two-state** (**binary** or **presence-absence**) characters (habit woody or herbaceous). However, all characters may not be two-state. They may be qualitative multistate (flowers white, red, blue) or quantitative multistate (leaves two, three, four, five at each node). Such multistate characters can be converted into two-state (flowers white or coloured; leaves four or more vs leaves less than four). Or else the characters may be split (flowers white vs not white, red vs not red, blue vs not blue; leaves two vs not two, three vs not three and so on). Such a splitting may, however, give more weightage to one original character (flower colour or number of leaves). The large number of data used will nevertheless neutralise any such bias. The two-state or binary characters are best coded as 0 and 1 for two alternate states. Such a coding is handled very efficiently by computers. The coded data may be entered in the form of a matrix with t number of rows (OTUs) and n number of columns (characters) with the dimension of the matrix (and the number of attributes) being $t \times n$ (Table 9.1).

Residual weighting involves excluding a character from the list when information for a large number of OTUs is not available, or is irrelevant. But in certain cases information may be available for a particular character for a large number of OTUs but not for a few. Alternately, the information may be irrelevant for a few OTUs (say number of spurs in an OTU, which lacks spurs). Such characters are used in analysis but for the OTUs for which information is not available or is irrelevant, an NC code (Not Comparable) is entered in the matrix.

Whenever the NC code is encountered the program bypasses that particular character for comparing the concerned OTU. For data handling by computers the NC code is assigned a particular (not 0 or 1) numeric value.

Although other forms of coding could also be handled, these are often more cumbersome. One solution is to give a separate symbol to each character state as explained below:

Flower Colour	Character state
White	A
Red	B
Yellow	C
Purple	D

Table 9.1: A portion of the data matrix with hypothetical *t* OTUs and *n* characters. Binary coding involves 0 for state *a* and 1 for state *b*. The NC code stands for characters not comparable for that OTU. In this analysis a total of 100 characters were used but only nine are pictured here.

OTUs (*t*)	Habit 0-woody 1-herbaceous	Fruit 0-follicle 1-achene	Ovary 0-superior 1-inferior	Leaves 0-simple 1-compound	Habitat 0-terrestrial 1-aquatic	Pollen 1-triporate 0-monosulcate	Ovule 1-unitegmic 0-bitegmic	Carpels 0-free 1-united	Plastids 1-PI-type 0-PII-type
1.	1	0	1	1	0	1	1	1	1
2.	1	0	1	1	1	1	0	0	1
3.	0	NC	0	1	0	0	1	1	1
4.	1	1	1	0	1	0	0	0	0
5.	1	0	1	1	0	1	1	0	1
6.	1	1	0	1	1	NC	0	1	0
7.	0	1	1	0	1	1	0	0	1
8.	0	0	0	1	0	0	1	1	1
9.	1	1	1	0	0	1	1	0	0
10.	0	0	0	1	1	0	1	1	1
11.	1	0	1	1	0	1	0	0	1
12.	1	1	0	0	1	1	0	0	1
13.	0	0	1	0	1	0	1	0	1
14.	1	0	1	0	1	0	1	0	1
15.	0	1	1	0	0	1	1	0	0

A match is scored if the same symbol occurs in two OTUs, otherwise a mismatch is recorded.

Measurement of Resemblance

Once the data have been codified and entered in the form of a matrix, the next step is to calculate the degree of resemblance between every pair of OTUs. A number of formulae have been proposed by various authors to calculate **similarity** or **dissimilarity (taxonomic distance)** between the OTUs. If we are calculating the similarity (or dissimilarity) based on binary data coded as 1 and 0, the following combinations are possible:

OTU k

	1	0
O 1	a	b
T		
U 0	c	d
j		

Number of matches $m = a + d$
Number of mismatches $u = b + c$
Sample size $n = a + b + c + d = m + u$
j and k are two OTUs under comparison

Some of the common formulae are discussed below:

1. Simple Matching Coefficient This measure of similarity is convenient and highly suitable for data wherein 0 and 1 represent two states of a character, and 0 does not merely represent the absence of a character state. The coefficient was introduced by Sokal and Michener (1958). The coefficient is represented as:

$$\text{\textit{Simple matching coefficient}} (S_{SM}) = \frac{\text{matches}}{\text{matches} + \text{mismatches}}$$

or
$$\frac{m}{m+u}$$

It is more convenient to record similarity in percentage. In that case the formula would read:

$$S_{SM} = \frac{m}{m+u} \times 100$$

When comparing a pair of OTUs, a match is scored when both OTUs show 1 or 0 for a particular character. On the other hand, if one OTU shows 0 and another 1 for a particular character, a mismatch is scored.

2. Coefficient of Association This coefficient was first developed by Jaccard (1908) and gives weightage to scores of 1 only. This formula is thus suitable for data where absence-presence is coded and 1 represents the presence of a particular character state, and 0 its absence. The formula is presented as:

$$\text{\textit{Coefficient of Association}} (S_J) = \frac{a}{a+u}$$

where a stands for number of characters that are present (scored 1) in both OTUs . This can similarly be represented as a percentage similarity.

3. Yule Coefficient This coefficient has been less commonly used in numerical taxonomy. It is calculated as:

$$S_Y = (ad - bc)/(ad + bc)$$

4. Taxonomic Distance Taxonomic distance between the OTUs can be easily calculated as a value against 100% similarity. It can also be directly calculated as **Euclidean distance** using the formula proposed by Sokal (1961):

$$\Delta jk = \left[\sum_{i=1}^{n} (X_{ij} - X_{ik})^2 \right]^{1/2}$$

The average distance would be represented as:

$$d_{jk} = \sqrt{\Delta^2_{jk}/n}$$

Other commonly used distance measures include **Mean character difference** (M.C.D.) proposed by Cain and Harrison (1958), **Manhattan metric distance coefficient** (Lance andWilliams, 1967) and **Coefficient of divergence** (Clark, 1952).

Once the similarity (or distance) between every pair of OTUs has been calculated the data are presented in a second matrix with $t \times t$ dimensions where both rows and columns represent OTUs (Table 9.2). It must be noted that diagonal t value in the matrix represents self-comparison of OTUs and thus 100% similarity. These values are redundant as such. The values in the triangle above this diagonal line would be similar to the triangle below. The effective number of similarity values as such would be $t \times (t-1)/2$. Thus as 15 OTUs are compared the number of values calculated would be $15 \times (15-1)/2 = 105$.

Table 9.2: Similarity matrix of the representative hypothetical taxa presented as percentage simple matching coefficient

→OTUs	1	2	3	4	5	6	7	8	9	10	11	12	13	14	15
1	100.0														
2	47.0	100.0													
3	54.0	47.0	100.0												
4	49.0	54.0	52.0	100.0											
5	50.0	51.0	44.0	49.0	100.0										
6	46.0	59.0	46.0	47.0	48.0	100.0									
7	47.0	48.0	48.0	46.0	65.0	47.0	100.0								
8	56.0	51.0	56.0	52.0	46.0	58.0	25.0	100.0							
9	50.0	45.0	49.0	50.0	60.0	40.0	79.0	30.0	100.0						
10	50.0	45.0	54.0	50.5	58.0	41.0	77.0	36.0	92.0	100.0					
11	53.0	54.0	49.0	45.5	65.0	51.0	92.0	31.0	75.0	73.0	100.0				
12	48.0	47.0	49.0	50.0	58.0	42.0	81.0	30.0	96.0	94.0	75.0	100.0			
13	47.0	44.0	49.0	49.5	59.0	44.0	68.0	41.0	81.0	83.0	62.0	81.0	100.0		
14	55.0	46.0	55.0	51.5	57.0	44.0	72.0	39.0	81.0	81.0	72.0	81.0	74.0	100.0	
15	56.0	45.0	57.0	53.0	54.0	44.0	67.0	40.0	78.0	72.0	67.0	74.0	67.0	87.0	100.0

Cluster Analysis

Data presented in OTUs × OTUs ($t \times t$) matrix are too exhaustive to provide any meaningful picture and need to be further condensed to enable a comparison of units. Cluster analysis is one such method in which OTUs are arranged in the order of decreasing similarity. The earlier methods of cluster analysis were cumbersome and involved shifting of cells with similar values in the matrix so that OTUs with closely similar similarity values were brought together as clusters. Today, with the advancement of computer technology, programs are available which can perform an efficient cluster analysis and help in the construction of cluster diagrams or phenograms. The various clustering procedures are classified under two categories.

1. Agglomerative Methods   Agglomerative methods start with *t* clusters equal to the number of OTUs. These are successively merged until a single cluster has finally been formed. The most commonly used clustering method in biology is the Sequential Agglomerative Hierarchic Non-overlapping clustering method (SAHN). The method is useful for achieving hierarchical classifications. The procedure starts with the assumption that only those OTUs would be merged which show 100% similarity. As no two OTUs would show 100% similarity we start with *t* number of clusters. Let us now lower the criterion for merger as 99% similarity; still no OTUs would be merged as in our example the highest similarity recorded is 96.0%. The best logical solution would be to pick up the highest similarity value (here 96.0) and merge the two concerned OTUs (here 9 and 12). By inference, if our criterion for merger is 96.0 we will have *t* − 1 clusters. Subsequently the next lower similarity value is picked up and the number of clusters reduced to *t* − 2. The procedure is continued until we are left with a single cluster at the lowest significant similarity value. Since at various steps of clustering a candidate OTU for merger would cluster with a group of OTUs, it is important to decide the value that would link the clusters horizontally in a cluster diagram. A number of strategies are used for the purpose.

In the commonly used **single linkage clustering** method (**nearest neighbour** technique or **minimum method**) the candidate OTU for admission to a cluster has similarity to that cluster equal to the similarity to the closest member within the cluster. The connections between OTUs and clusters and between two clusters are established by single links between pairs of OTUs. This procedure frequently leads to long straggly clusters in comparison with other SAHN cluster methods. The phenogram for our data using this strategy is shown in Fig. 9.2. The highest similarity value in our matrix (see Table 9.2) is 96.0 between OTUs 9 and 12, and as such they are linked at that level. The next similarity value of 94.0 is between OTUs 10 and 12, but since 12 has already been clustered with 9, 10 will join this cluster linked at 94.0. The process is repeated until all OTUs have been agglomerated into a single cluster at similarity value of 53.0.

In the **complete linkage clustering** method (**farthest neighbour** or **maximum method**) the candidate OTU for admission to a cluster has similarity to that cluster equal to its similarity to the farthest member within the cluster. This method will generally lead to tight discrete clusters that join others only with difficulty and at relatively low overall similarity values.

In the **average linkage clustering** method an average of similarity is calculated between a candidate OTU and a cluster or between two clusters. Several variations of this average method are used. The unweighted pair-group method using arithmetic averages (UPGMA) computes the average

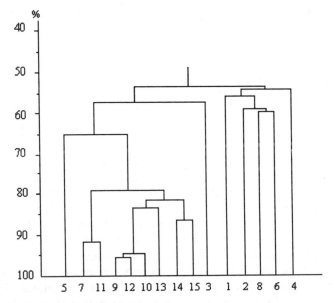

Fig. 9.2 Cluster diagram of 15 OTUs based on similarity matrix in Table 9.2 using single linkage strategy.

similarity or dissimilarity of a candidate OTU to a cluster, weighting each OTU in the cluster equally regardless of its structural subdivision. The weighted pair-group centroid method (WPGMC) weighs the most recently admitted OTU in a cluster equally to the previous members of the cluster.

2. Divisive methods   Divisive methods, as opposed to agglomerative methods, start, with all *t* OTUs as a single set, subdividing this into one or more subsets; this is continued until further subdivision is not necessary. The commonly used divisive method is **association analysis** (William, Lambert and Lance, 1966). The method has been mostly used in ecological data employing two state characters. It builds a dendrogram from the top downwards as opposed to cluster analysis which builds a diagram from the bottom up. The first step in the analysis involves calculating **chi square** value between every pair of characters using the formula:

$$X^2_{hi} = n(ad - bc)^2/[(a + b)(a + c)(b + d)(c + d)]$$

where *i* stands for the character being compared and *h* for any character other than *i*. For each character the sum of chi square is computed and the character showing maximum **chi square** value is chosen as the first differentiating character. The whole set of OTUs is divided into two clusters, one containing the OTUs which show character state *a* and another containing OTUs which show character state *b*. Within each cluster again the character with the next value of the sum of chi square is selected and the cluster subdivided into two clusters as before. The process is repeated until further subdivision is not significant.

Hierarchical Classifications

The phenogram constructed using any technique or strategy can be used for attempting hierarchical classification, by deciding about certain threshold levels for different ranks. One may tentatively decide 85% similarity as the threshold for the species, 65 for genera and 45 for families and recognise these ranks on the basis of number of clusters recognised at that threshold. Whereas such an assumption can help in hierarchical classification, the point of conflict would always be the threshold level for a particular rank. Some may argue, and are justified in doing so, to suggest 80% (or any other value) as the threshold for species. It is more common, therefore, to use the terms **85% phenon line, 65% phenon line, and 45% phenon line**. These terms may conveniently be used until such time that sufficient data are available to assign formal taxonomic ranks to the various phenon lines.

The results of cluster analysis are commonly presented as dendrograms known as **phenograms**. They can also be presented as **contour diagrams** (Fig. 9.3), originally developed under the name **Wroclaw diagram** by Polish phytosociologists. The contour diagram may also incorporate the levels at which clustering has taken place.

Ordination

Ordination is a technique which determines the placement of OTUs in two-dimensional or three-dimensional space. The results of two-dimensional ordination are conveniently represented with the help of a scatter diagram and those of three-dimensional ordination with the help of a three-dimensional model. The procedure works on distance values calculated directly from the coded data or indirectly from already calculated similarity values as 100 minus similarity (if similarity values are in percentage) or 1 minus similarity (if similarity values range between 0 to 1). A dissimilarity matrix based on Table 9.2 is shown in Table 9.3. The first step in ordination starts with construction of the x (horizontal) axis. In the commonly used method of **polar ordination** the two most distant OTUs are selected as the end points (A and B) on the x-axis. In our example these are OTU 8 and 7 with a distance (dissimilarity value) of 75. The position of all other OTUs on this axis can be plotted one by one. OTU 10 has a distance of 64 from A (OTU 8) and a distance of 23 from B (OTU 7). A compass with a radius of 64 units is swung from A and a compass with a radius of 23 units is swung from B, forming two arcs. A line joining the intersection of two arcs forms a perpendicular on the x-axis, and the point at which the line crosses the x-axis is the position of the OTU. The distance between the x-axis and the point of intersection of arcs is the **poorness of fit** of the concerned OTU. The location of

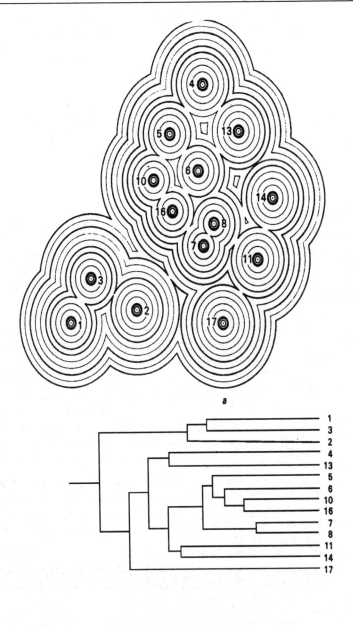

Fig. 9.3   Contour diagram based on the phenogram shown alongside.

OTU on the axis from the left (point A) can also be calculated directly instead of plotting:

$$x = \frac{L^2 + dAC^2 - dBC^2}{2L}$$

where x is the distance from the left end, L is the dissimilarity value between A and B (length of x-axis), dAC is dissimilarity between A and the OTU under consideration and dBC the dissimilarity between B and the OTU under consideration. The poorness of fit (e) of this OTU can be calculated as:

$$e = \sqrt{dAC^2 - x^2}$$

After the position of all OTUs has been determined and the poorness of fit calculated, a second axis (vertical axis or y-axis) has to be constructed. For this the OTU with highest poorness of fit (most poorly fitted to x-axis) is selected and this forms the first reference OTU of the y-axis. The second reference stand is selected as that one with the highest dissimilarity to the first reference OTU of y-axis, but within 10% (of the length of x-axis) distance on x-axis. The position of all other OTUs on the y-axis and their

Table 9.3 Dissimilarity matrix of the representative hypothetical taxa based on the similarity matrix in Table 9.2

OTUs	1	2	3	4	5	6	7	8	9	10	11	12	13	14	15
1	0.0														
2	53.0	0.0													
3	46.0	53.0	0.0												
4	51.0	46.0	48.0	0.0											
5	50.0	49.0	56.0	51.0	0.0										
6	54.0	41.0	54.0	53.0	52.0	0.0									
7	53.0	52.0	52.0	54.0	35.0	53.0	0.0								
8	44.0	49.0	44.0	48.0	54.0	42.0	75.0	0.0							
9	50.0	55.0	51.0	50.0	40.0	60.0	21.0	70.0	0.0						
10	50.0	55.0	46.0	49.5	42.0	59.0	23.0	64.0	8.0	0.0					
11	47.0	46.0	51.0	54.5	35.0	49.0	8.0	69.0	25.0	27.0	0.0				
12	52.0	53.0	51.0	50.0	42.0	58.0	19.0	70.0	4.0	6.0	25.0	0.0			
13	53.0	56.0	51.0	50.5	41.0	56.0	32.0	59.0	19.0	17.0	38.0	19.0	0.0		
14	45.0	54.0	45.0	48.5	43.0	56.0	28.0	61.0	19.0	19.0	28.0	19.0	26.0	0.0	
15	44.0	55.0	43.0	47.0	46.0	56.0	33.0	60.0	22.0	28.0	33.0	26.0	33.0	13.0	0.0

poorness of fit is determined as earlier. By using the values of poorness of fit to y-axis a z-axis can similarly be generated and the position of all OTUs on z-axis determined similarly. The values can be used for constructing a **scatter diagram** or a three-dimensional model.

A commonly used ordination technique known as **principal component analysis** similarly calculates values for a two-dimensional scatter diagram. In this method, however, the values on the horizontal as well as the vertical axis are non-zero, ranging from –1 to 1 (calculated as **eigenvalues**) and as such the scatter diagram is presented along four axes: positive horizontal, negative horizontal, positive vertical and negative vertical (Fig 9.4). The technique is based on the assumption that if a straight line represented a single character, all the OTUs could be placed along the line according to their value for that character. If two characters were used, a two-dimensional graph would suffice to locate all OTUs. With n characters, n-dimensional space is required to locate all OTUs as points in space. Principal component analysis determines the line through the cloud of points that accounts for the greatest amount of variation. This is the first principal component axis. A second axis produced perpendicular to the first, accounts for the next greatest amount of variation. The procedure ultimately produces axes one less than the number of OTUs. The first two axes are generally plotted to produce a scatter diagram. The procedure also calculates **eigenvectors**, which indicate the importance of a character to a particular axis. The larger the eigenvector in absolute value, the more important that character.

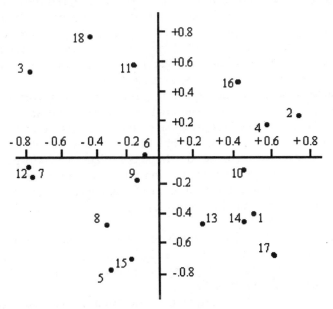

Fig. 9.4 Plot of the results of the principal component analysis of 18 hypothetical taxa.

A related method of ordination is **principal co-ordinate analysis** developed by Gower (1966). This technique enables computation of principal components of any Euclidean distance matrix without being in possession of original data matrix. The method is also applicable to non-Euclidean distance and association coefficients as long as the matrix has no large negative eigenvalues. Principal co-ordinate analysis also seems to be less disturbed by NC entries than principal components.

Application of Numerical Taxonomy in Angiosperms

There are several examples of the utilisation of numerical analysis in solving taxomomic problems. Clifford (1977) performed numerical analysis of four subclasses of monocotyledons recognised by Takhtajan (1969) and Cronquist (1968), namely Alismidae (Cronquist Alismatidae), Liliidae, Commelinidae and Arecidae. On the basis of the results of cluster analysis and principal co-ordinate analysis he concluded that water lilies (Alismidae) form a distinct group among monocotyledons. The study also concluded that, Triuridales (placed by Cronquist in Alismatidae and by Takhtajan in Liliidae) belong to a distinct group among wind-pollinated families. It is interesting to note that Takhtajan (1987, 1997) separated Triuridales under a distinct subclass Triurididae, and Thorne (1992) also placed the order under a distinct superorder, Triuridanae. The study supported placement of Typhales among Arecidae by Takhtajan but opposed the placement of Arales. The study concluded that although Arales are not related to Arecidae they sit uneasily even among 'lilies' (Liliidae) to which they are closer. It is significant that Takhtajan (1997) separated Arales to a distinct subclass, Aridae. Cluster analysis yielded a dendrogram (phenogram) with 10 groups. The results of principal co-ordinate analysis revealed that attributes most closely correlated with the first principal co-ordinate axis were the presence or absence of squamulae, the nature of anther tapetum and the presence or absence of endosperm in the seed. The squamulae and endosperm were highly diagnostic of the primary division within the monocots. The four attributes most diagnostic for separation of the wind-pollinated group of families from the animal-pollinated groups included ovary, endosperm, subsidiary cells and fruit.

Another significant work on dicotyledons was undertaken by Young and Watson (1971) who used 83 attributes in their analysis of 543 representative genera. On the basis of their study they proposed the division of dicots into Crassinucellatae and Tenuinucellatae.

Hilu and Wright (1982) undertook numerical analysis of Gramineae. Utilisation of morphological and microscopic characters separately did not give satisfactory results, as the results were incongruent, but a combination of both resulted in the establishment of eight clusters comparable to the eight subfamilies recognised within the family.

II. PHYLOGENETIC METHODS: CLADISTICS

Phylogenetic methods aim at developing a classification based on an analysis of phylogenetic data, and developing a diagram termed a **cladogram**, which depicts the genealogical descent of taxa. Biologists who practice this methodology are known as **cladists**. Although phylogenetic diagrams (now appropriately known as **phylograms**) have been used by Bessey (1915), Hutchinson (1959, 1973), and contemporary authors of classification systems to show relationships between taxa, the cladograms are distinct in that they are developed using a distinct methodology. This method was first proposed by W. Hennig (1950, 1957), a German zoologist who founded the subject of **phylogenetic systematics**. The term **cladistics** for this methodology was coined by Mayr (1969). An American Botanist, W.H. Wagner, working independently developed a method of constructing phylogenetic trees, called the groundplan-divergence method, in 1948. Over the years cladistics has developed into a forceful methodology for developing phylogenetic classifications.

Cladistics is a methodology that attempts to analyse phylogenetic data objectively, in a manner parallel to taxometrics which analyses phenetic data. Cladistic methods are largely based on the principle of parsimony according to which the most likely evolutionary route is the shortest hypothetical pathway of changes that explains the pattern under observation. Taxa in a truly phylogenetic system should be monophyletic. It has been found that **symplesiomorphy** (possession of primitive or plesiomorphic character-state in common by two or more taxa) does not necessarily indicate monophyly. **Synapomorphy** (possession of derived or apomorphic character-state in common by two or more taxa), on the other hand, is a more reliable indicative of monophyly. It is thus common to use homologous shared and derived characters for cladistic studies. The methodology of cladistics is similar to taxometrics in several aspects.

Operational Evolutionary Units

The working units in cladistics are termed Operational Evolutionary Units (OEUs or simply EUs), equal to and often the same as OTUs of taxometrics. A unique feature of cladistic studies, however, is that the list of OEUs generally includes a hypothetical ancestor, comparison with which reveals crucial phylogenetic information.

Characters and Coding

A list of characters is prepared next and the plesiomorphic and apomorphic states of each character distinguished. It is important to distinguish plesiomorphic and apomorphic states of a character before an analysis is undertaken. A preliminary step as such involves a **character compatibility** study in which each character is examined to determine the proper sequence of character-state changes that take place as the evolution progresses. Whereas the majority of characters included in analysis present two character-states, in some cases more than two character-states may be encountered. In such cases more than one **morphoclines (transformation series)** are possible as illustrated below:

Character-states	Morphoclines possible
0 and 1	$0 \leftrightarrow 1$
0, 1 and 2	$0 \leftrightarrow 1 \leftrightarrow 2$
	$1 \leftrightarrow 0 \leftrightarrow 2$
	$0 \leftrightarrow 2 \leftrightarrow 1$

It is thus necessary to determine the relative ancestry of the character-states, or the assignment of **polarity**. The designation of polarity is often one of the more difficult and uncertain aspects of phylogenetic analysis. For this the comparison may be made within the concerned group (**in-group comparison**) or relatives outside the group (**out-group comparison**). The latter may often provide useful information especially when the out-group used is the **sister-group** of the concerned group. If two character-states of a character are found in a single monophyletic group, the state that is also found in a sister-group is likely to be plesiomorphic and that found only within the concerned monophyletic group is likely to be apomorphic. **In-group comparison** (also known as **common ground plan** or **commonality principle**) is based on the presumption that in a given group (presumably monophyletic), the primitive structure would tend to be more common. It is assumed that the evolution of a derived condition will occur in only one of potentially numerous lineages of the group; thus the ancestral condition will tend to be in the majority. Once a set of data relating to plesiomorphic and apomorphic character-states has been accumulated for all EUs a data matrix t (EUs) $\times$ n (characters) is prepared, in which 0 represents the plesiomorphic character-state and 1 the apomorphic character-state of a particular character. The list of EUs in the matrix also includes a hypothetical ancestor as the last row in the matrix (Table 9.3), the wherein all 0 character-states are scored. In cases where the character encounters three character states (leaf simple → pinnately lobed → pinnately compound) the coding may accordingly be done as 0 for the most primitive character-state (simple leaf), 1 for the intermediate character-state (pinnately lobed leaf) and 2 for the most advanced state (pinnately compound leaf).

Measure of Distance

A data matrix with coded character-states for each EU can be used for calculating the distance between every pair of EUs, including the hypothetical ancestor. The distance is calculated as the total number of

Table 9.3 Data matrix of *t* EUs and *n* characters scored as 0 (plesiomorphic) and 1 (apomorphic) character-states. The matrix is similar to Table 9.1 but only 9 characters pictured are used for calculations. Also the last EU included is the hypothetical ancestor in which all character-states are scored as 0 (plesiomorphic), as it is presumed that the ancestor would possess all characters in a plesiomorphic state.

Characters (n) → EUs (t)	Habit 0-woody 1-herbaceous	Fruit 0-follicle 1-achene	Ovary 0-superior 1-inferior	Leaves 0-simple 1-compound	Habitat 0-terrestrial 1-aquatic	Pollen 1-triporate 0-monosulcate	Ovule 1-unitegmic 0-bitegmic	Carpels 0-free 1-united	Plastids 1-PI-type 0-PII-type
1.	1	0	1	1	0	1	1	1	1
2.	1	0	1	1	1	1	0	0	1
3.	0	1	0	1	0	0	1	1	1
4.	1	1	1	0	1	0	0	0	0
5.	1	0	1	1	0	1	1	0	1
6.	1	1	0	1	1	1	0	1	0
7.	0	1	1	0	1	1	0	0	1
8.	0	0	0	1	0	0	1	1	1
9.	1	1	1	0	0	1	1	0	0
10.	0	0	0	1	1	0	1	1	1
11.	1	0	1	1	0	1	0	0	1
12.	1	1	0	0	1	1	0	0	1
13.	0	0	1	0	1	0	1	0	1
14.	1	0	1	0	1	0	1	0	1
15.	0	0	0	0	0	0	0	0	0

character-state differences between two concerned EUs, often represented as Manhattan distance and the data presented as *t* × *t* matrix (Table 9.4). The distance can be calculated as under:

$$d(X, Y) = \sum_{i=1}^{n} |V_{Xi} - V_{Yi}|$$

where $d(X,Y)$ is the distance between taxa X and Y, n the total number of characters, V_{Xi} the character-state value of EU X for character i and V_{Yi} the character-state value of EU Y for character i. This method is closer to taxometric methods because both plesiomorphic and apomorphic character-states are given equal weightage.

Another method of calculating distance involves calculation of the number of apomorphic character-states common between the pairs of concerned EUs, ignoring the possession of plesiomorphic character-states in common (Table 9.5). Since only synapomorphy is likely to define monophyletic groups, this method is closer to the original cladistic concept.

Table 9.4 $t \times t$ matrix presenting distance between EUs expressed as number of character-state differences between pairs of EUs

↓EUs	1	2	3	4	5	6	7	8	9	10	11	12	13	14	15
1	0														
2	3	0													
3	4	7	0												
4	7	4	8	0											
5	1	2	5	6	0										
6	6	5	6	4	7	0									
7	6	3	7	3	5	6	0								
8	3	6	1	8	4	6	7	0							
9	4	5	7	3	3	6	4	7	0						
10	4	5	2	7	5	5	6	1	8	0					
11	2	1	6	5	1	6	4	5	4	6	0				
12	6	3	7	3	5	4	2	7	4	6	4	0			
13	5	4	5	4	4	8	3	4	5	3	5	5	0		
14	4	3	6	3	3	7	4	5	4	4	4	4	1	0	
15	5	6	6	4	4	7	3	6	1	7	5	5	4	5	

Table 9.5 $t \times t$ matrix presenting distance between EUs expressed as number of derived (apomorphic) character-states common between pairs of EUs

↓EUs	1	2	3	4	5	6	7	8	9	10	11	12	13	14	15
1	×														
2	5	×													
3	4	2	×												
4	2	3	0	×											
5	6	5	3	2	×										
6	3	3	2	3	2	×									
7	3	4	1	3	3	2	×								
8	4	2	4	0	3	2	1	×							
9	4	3	1	3	4	2	3	1	×						
10	4	3	4	1	3	3	2	4	1	×					
11	5	5	2	2	5	2	3	2	3	2	×				
12	3	4	1	3	3	3	4	1	3	2	3	×			
13	3	3	2	2	3	1	3	2	2	3	2	2	×		
14	4	4	2	3	4	2	3	2	3	3	3	3	4	×	
15	3	2	1	2	3	1	3	1	4	1	2	2	2	2	×

Construction of Cladograms

Different methods are available for the final analysis of cladistic information. In the **rooted trees** method a hypothetical ancestral taxon constitutes the basal member of the tree. The evolutionary polarity of taxa is decided for construction of such trees. The **Wagner groundplan divergence method**: an example of this was first developed by H.W. Wagner in 1948 as a technique for determining the phylogenetic relationships among organisms that he hoped would replace intuition with analysis. The method was based on determining the apomorphic character-states present within a taxon and then linking the subtaxa based on relative degree of apomorphy. Interestingly, whereas the method found little favour with zoologists, it has been used in many botanical studies. The following steps are involved in the analysis:

1. Determine which of the various characters (or character-states) in a series of character transformations are apomorphic.

2. Assign the score of 0 to the plesiomorphic character and 1 to the apomorphic character in each transformation series. If the transformation series contains more than two homologues then these 'intermediate apomorphies' may be scaled between 0 and 1. Thus in a transformation series three characters may be scored as 0, 0.5 and 1.

3. Construct a table of taxa (EUs) and coded characters (or character-states: see Table 9.3).

4. Determine the **divergence index** for each taxon by totalling the values. Since apomorphic character-states are coded 1, the divergence index in effect represents the number of apomorphies (character-states) in a taxon. For the data matrix in Table 9.3 the divergence index for 15 taxa would be calculated as:

Taxon	Divergence index
1	7
2	6
3	5
4	4
5	6
6	6
7	5
8	4
9	5
10	5
11	5
12	5
13	4
14	5
15	0

Note that the hypothetical ancestral taxon 15 has an index of 0.

5. Plot the taxa on a graph, placing each taxon on a concentric semicircle that equals its divergence index. The lines connecting the taxa are determined by shared synapomorphies (see Table 9.5). The cladogram **(Wagner tree)** is presented in Fig. 9.5.

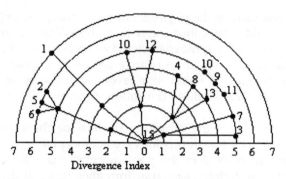

Fig. 9.5 General representation of a Wagner tree.

Kluge and Farris (1969) developed a computer algorithm for computing Wagner trees, based on the principle of parsimony.

The second type of cladograms are **unrooted trees** or **networks**. Such cladograms (Fig. 9.6) are non-directional and no polarity of characters is specified. These unrooted trees may be converted into rooted trees, however, by an *a posteriori* decision about the most primitive end of the network. Such decisions are based on the combination of characters possessed by the various taxa and not on individual characters.

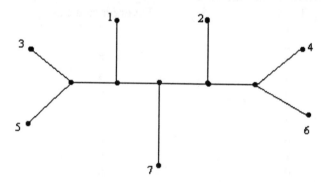

Fig. 9.6 Representative unrooted tree

Cladistic analysis is primarily based on the assumption that evolution is an ordered, divergent step-wise transformation of characters from plesiomorphic to apomorphic states. There may, however, be situations of convergence, parallelism or even reversal of characters, which if undetected may result in distorted cladograms.

Not all cladistic methods apply the principle of parsimony. The methods of **compatibility analysis** or **clique analysis** utilise the concept of character compatibility. Such methods can detect and thus omit homoplasy. They can be carried out manually or using a computer program and can generate both rooted as well as unrooted trees. Groups of mutually compatible characters are termed **cliques**. Let us consider two characters, A and B, with two

	A1	A2
B1	A1B1	A2B1
B2	A1B2	A2B2

character-states each. Four character-state combinations are possible:

Assuming the evolution has proceeded from A1 to A2 and from B1 to B2, if all the four combinations are met in nature then obviously there must have been at least one reversal (A2 to A1) or parallelism (A1 to A2 occurring twice), and as such A and B are incompatible. On the other hand, if only two or three of the combinations occur, then A and B are compatible. Cliques are formed by comparing all pairs of characters and finding mutually compatible sets. The largest clique is selected from the data to produce a cladogram. Finally, a rooted tree or network is obtained according to whether or not a hypothetical ancestor was included in the analysis.

10

Variation and Speciation

It is now universally agreed that different species are not fixed entities but systems of populations which show variation and wherein no two individuals are identical. This concept of variations was first proposed by Lamarck and further developed by Darwin, culminating in his famous book *Origin of Species* (1859). Systematics is a unique natural science concerned with the study of individual, population and taxon relationships for purposes of classification. The study of plant systematics is based on the premise that in the tremendous variation in the plant world, there exist conceptual discrete units (usually named as species) that can be recognised, classified, described, and named, and on the further premise that logical relationships developed through evolution exist among these units.

I. TYPES OF VARIATION

The recognition of taxonomic units is based on identification of the occurrence and the degree of discontinuity in variation in the populations. The variation may be **continuous** when the individuals of a population are separable by infinitely small differences in any of the attributes. In **discontinuous variation**, however, there is a distinct gap between two populations, each showing its own continuous variation for a particular attribute. The discontinuity between the populations primarily results from **isolation** in nature. Isolation plays a major role in establishing and widening the gap between the populations, allowing evolution to take its destined course with no disturbance. Variation in plants includes three fundamental types: developmental, environmental and genetic.

Developmental Variation

A distinct change in attributes is often found during different stages of development. Juvenile leaves of *Eucalyptus, Salix* and *Populus* are often different from mature leaves, and may often cause much confusion, but may prove equally useful when both types of leaves are available from a plant. The first leaves of *Phaseolus* are opposite and simple, the later ones alternate and pinnately compound. Because the seedling stage is most critical in a plant's life, the characters present during this period surely have survival value. Takhtajan proposed a **neotenous** origin for angiosperms on the assumption of juvenile simple leaves of seed ferns having persisted in the adult forms, which were the direct progenitors of angiosperms.

Environmental Variation

Environmental factors often play a very major role in shaping the appearance of a plant. Heterophylly is the common manifestation of environmental variation. The submerged leaves of *Ranunuculus aquatilis* are finely dissected whereas the emergent leaves of the same plant are broadly lobed. The first submerged leaves of *Sium suave* are pinnately dissected and flaccid, the older emerged leaves are pinnately compound and stiff. The individuals of a species often exhibit **phenotypic plasticity**, expressing diferent phenotypes under different environmental conditions. Such populations are named **ecophenes**. In *Epilobium* sun-plants have small, thick leaves, many hairs and a short stature, whereas shade-plants have larger thinner leaves with fewer hairs and a taller stature.

Genetic Variation

Genetic variation may result from **mutation** or **recombination**. Mutation is the occurrence of heritable change in the genotype of an organism that was not inherited from its ancestors. It is the ultimate source of variation in a species and replenishes the supply of genetic variability. A mutation may be as minute as the substitution of a single nucleotide pair in the DNA (**point mutation**) or as great as major change in chromosome structure (**chromosomal mutation**). Chromosomal mutation may be due to deletion, inversion, aneuploidy or polyploidy. Recombination is a reassortment of chromosomes, bringing together via meiosis and fertilisation the genetic material from different parents and producing a new genotype.

II. VARIANCE ANALYSIS

Since no two individuals in a population are similar there is need for some objective analysis for useful comparison. The simplest tool is to calculate **mean** by adding the series of values and dividing the total by the number of values. The formula for calculating mean is:

$$\overline{X} = \frac{\Sigma X_i}{n}$$

where $\overline{X}$ represents the mean, Σ summation of all values of X, X_i represents the individual values of an attribute under study and n represents the number of values. Thus five plants of a species with height 15 cm, 12 cm, 10 cm, 22 cm and 16 cm would have a mean of 15 cm ((15+12+10+22+16)/5). The extent of variation within a population of a species is best represented by determining **variance**. If various individuals were not far from this mean the variance would be minimum. On the other hand if many individuals were far removed from mean, the variance would be higher. The variance may either be calculated for a population, or a sample from the population. The variance for a population may be calculated as:

$$\sigma^2 = \frac{\Sigma(X_i - \overline{X})^2}{n}$$

To obtain variance the difference between each value of the attribute (X) and the mean is squared and a sum of these squares is divided by the number of observation (n). For calculating sample variance (s^2) the sum of squares is divided by $n - 1$ instead of n. The formula for sample variance may be written as:

$$s^2 = \frac{\Sigma(X_i - \overline{X})^2}{n-1}$$

The square root of variance is represented by **standard deviation**. We may thus determine standard deviation of a population as:

$$\sigma = \sqrt{\sigma^2}$$

and that for a sample as:

$$s = \sqrt{s^2}$$

For our sample data the sample variance would be $[(15 - 15)^2 + (12 - 15)^2 + (10 - 15)^2 + (22 - 15)^2 + (16 - 15)^2]/4 = 21$ and the sample standard deviation

$$\sqrt{21} = 4.5825$$

III. ISOLATING MECHANISMS

Isolation is the key factor preventing intermingling of distinct species through prevention of hybridisation. Based on whether isolating mechanisms operate before or after sexual fusion, two main types of mechanisms are distinguished: **prezygotic mechanisms** and **postzygotic mechanisms**. A detailed classification of isolating mechanisms is presented below.

A. Prezygotic Mechanisms (operating before sexual fusion)

I. *Pre-pollination mechanisms*
   1. *Geographical isolation*: Two species are separated geographically by a gap larger than their pollen and seed dispersal. *Platanus orientalis* (Mediterranean region) and *P. occidentalis* (North America) are well-

established species but readily interbreed when brought into the same area (vicarious species).

2. *Ecological isolation*: Two species occupy the same general area but occupy different habitats. *Silene alba* grows on light soils in open places and *S. dioica* on heavy soils in shade. Their habitats rarely overlap, but when they do, hybrids are encountered.

3. *Seasonal isolation*: Two species occur in the same region but flower at different seasons. *Sambucus racemosa* and *S. nigra* flower nearly 7 weeks apart.

4. *Temporal isolation*: Two species flower during the same period but during different times of the day. *Agrostis tenuis* flowers in the afternoon, whereas *A. stolonifera* flowers in the morning.

5. *Ethological isolation*: Two species are interfertile but have different pollinators. Hummingbirds for example, are attracted to red flowers and hawk-moths to white flowers.

6. *Mechanical isolation*: Pollination between two related species is prevented by structural differences between flowers, as for example between *Ophrys insectifera* and *O. apifera*.

II. *Post-pollination mechanisms*

1. *Gametophytic isolation*: This is the commonest isolating mechanism wherein cross-pollination occurs but the pollen tube fails to germinate or if germinated, the male gametes cannot reach the egg within the embryo sac.

2. *Gametic isolation:* In such cases, reported in several crop plants, the pollen tube releases the male gametes into the embryo sac, but gametic and/or endospermic fusion does not occur.

B. Postzygotic Mechanisms (operating after sexual fusion)

1. *Seed incompatibility*: The zygote or even immature embryo is formed but fails to develop and as such a mature seed is not formed. The phenomenon is commonly encountered in crosses between *Primula elatior* and *P. veris*.

2. *Hybrid inviability*: Mature seed is formed and manages to germinate but the F1 hybrid dies before the flowering stage is reached. The phenomenon is commonly encountered in crosses between *Papaver dubium* and *P. rhoeas*.

3. *F1 hybrid sterility*: F1 hybrids are fully viable and reach the flowering stage but flowers may abort or abortion may occur as late as F2 embryo formation, with the result that the F1 hybrid fails to produce viable seeds.

4. *F2 hybrid inviability or sterility*: F2 hybrid dies much before reaching the flowering stage or fails to produce seeds.

IV. SPECIATION

Speciation is a general term for a number of different processes which involve the production of new species. New species may develop through the mechanism of **abrupt speciation** or **gradual speciation**. The phenomenon of abrupt speciation (example of **sympatric speciation**) is commonly met in genera such as *Tragopogon* and *Senecio*. The species are often well isolated and any chance hybridisation fails to culminate in successful hybrids because of genomic differences. In some cases, however, hybridisation may be accompanied by chromosome duplication resulting in the formation of allopolyploids. Such allopolyploids depict normal pairing at meiosis and thus represent well-isolated, phenotypically as well as genotypically distinct species.

Gradual Speciation

This is a more common phenomenon in nature. It may involve **phyletic evolution** when one species might evolve into something different from its ancestor over a period of time (**phyletic speciation**). Alternatively, a population belonging to a single species might differentiate into two evolutionary lines through **divergent evolution (additive speciation)**.

Phyletic speciation The concept of phyletic speciation has been the subject of considerable debate. It is sequential production of species within a single evolutionary lineage. Species A might over a period of time change through species B and C into species D without ever splitting. The new species produced in fusion are variously called **successional species**, **palaeospecies** and **allochronic species**. The species which have become extinct in the process are termed **taxonomic extinctions**. Wiley (1981), while agreeing with the concept of phyletic character transformation, rejects the concept of phyletic speciation on the grounds that:
  (1) Recognition of phyletic species is an arbitrary practice. Mayr (1942) argues that delimitation of species which do not belong to the same time-scale is difficult.
  (2) Arbitrary species result in arbitrary speciation mechanisms.
  (3) Phyletic speciation has never been satisfactorily demonstrated.

Additive speciation Additive speciation is the commonest mode of speciation, which adds to the diversity of living organisms. Mayr (1963) suggested the occurrence of **reductive speciation**, whereby two previously independent species fuse into a third, new species, themselves becoming extinct. Hybridisation likewise produces new species but this always leads to addition in the number of species. It is impossible to imagine that two evolutionary species can actually fuse to produce a third species and themselves become extinct. This may happen in a particular region, but not over the entire range of these species. The various modes of additive speciation are described below:

1. *Allopatric speciation*: Lineage independence and consequent speciation result from geographical separation of lineages, i.e., the actual physical separation of two relatively large populations of a single species. Over a period of time such separation would enable geographical races to develop and maintain gene combinations controlling their morphological and physiological characters. The development of reproductive isolation would sooner or later result in the establishment of distinct species (Fig. 10.1).

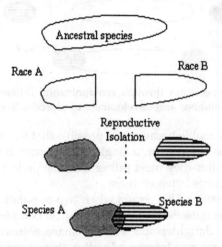

Fig. 10.1 Allopatric speciation resulting from geographical separation of populations of an ancestral species.

Allopatric speciation may also result from the development of new species along the boundaries of a large central population. These marginal populations (races) get separated from the main population when environmental differentiation occurs. They then undergo adaptive radiations to develop physical and physiological differences, which sooner or later get genetically fixed (ecotypes). With further morphological and physiological differentiation they form distinct varieties (or subspecies). Development of reproductive isolation establishes these as distinct species, that will retain their identity even if future chance should draw them together (Fig. 10.2).

2. *Allo-parapatric speciation*: Such speciation occurs when two populations of an ancestral species are separated, differentiate to a degree that is not sufficient for lineage independence, and then develop lineage independence during a period of parapatry (limited sympatry). It differs from allopatric speciation in that speciation is completed after a period of sympatry and the process of attaining lineage independence is

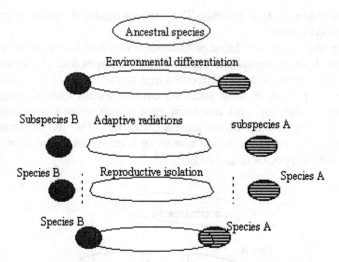

Fig. 10.2 Allopatric speciation through environmental differentiation, successive adaptive radiations and development of reproductive isolation.

potentially reversible because it is possible that two partly differentiated populations could form a single evolutionary lineage showing clinal variation after they meet rather than the period of sympatry reinforcing differences between them.

3. *Parapatric speciation*: This occurs when two populations of an ancestral species differentiate despite the fact that no complete disjunction has occurred. The daughter species may share a small fraction of their respective ranges and interbreed within this narrow contact zone and yet still differentiate.

4. *Stasipatric speciation*: This is similar to parapatric speciation except that it results from spontaneous chromosomal modifications. The resultant chromosome arrangement must be fully viable in the homozygous state but of reduced viability in the heterozygous state.

5. *Sympatric speciation*: This results in the production of new species with no geographical separation of populations, eventhough most cases of sympatric speciation, such as those resulting from hybridisation and apomixis belong to abrupt speciation. The process of ecological sympatric speciation is a slow one of gradual speciation. The ecological differences in the habitats result in adaptive radiations in populations which gradually evolve into new species.

11

Taxonomy on the Web

Internet revolution in recent years has seen an all-round development in scientific interaction, with instant access to huge databases around the world. There have been co-operative efforts to bring together botanical literature in general and taxonomic literature in particular, with interconnected links for convenient jumping around to various electronic sites on the web. The Botanical Museum of the Finnish Museum of Natural History, Helsinki maintains an **Internet Directory for Botany: Journals, Books, Literature Databases (IDB)**, which provides information under the headings: **General, Books, Journals CD, Literature databases,** and **Publishers and Booksellers**. This main page lists more important sites. In addition there are direct links to the following main topics which provide detailed listings.

1. **Latest Changes.** Atlas Florae Europeae (AFE) Database. Distributional maps published in *Atlas Florae Europeae* 1–11 can be viewed through a free evaluation version (for Windows 95 and NT) of the program.

2. **Search engines.** This section provides important guidelines for searching the botanical literature.

3. **Arboreta and Botanical Gardens.** Exhaustive list of Arboreta and Botanical Gardens a large number of which have their own servers and are inline for a large amount of information.

4. **Biologists' addresses**. A search engine maintained by BIOSCI, which is a set of electronic communication forums currently supported in the US by a grant from DOE with equipment from NSF. The UK BIOSCI node is supported by the Daresbury Laboratory. The World Taxonomist's Database is maintained by the Expert Centre for Taxonomic Identification (ETI), the Netherlands.

5. **Botanical Museums, Herbaria, Natural History.** Separate listings for these headings with easy links for inline access to all major locations throughout the world, with wide coverage of sites in the USA Updated information for US herbaria listed in the *Index Herbariorum* and its supplements (published in *Taxon*) is now available for searching by

institution, city, state, acronym, staff member, correspondent, and re-search speciality at the New York Botanical Garden Website (http://www.nybg.org/bsci/ih/ih.html). Telephone and fax numbers and e-mail and URL addresses are included.

6. **Botanical Societies, International Botanical Organisations.** The listing is included under these topics: Agriculture, Economic Botany, Forestry; Bryology; Ecology; Horticulture, Garden Plants (general); One genus/family/group societies; Mycology; National/Regional botanical and natural history societies; Palynology; Pathology; Phycology; Physiology; Systematics and taxonomy; Others.

7. **Checklists, Floras, Taxonomic Databases, Vegetation.** Information is provided under geographical headings: General, Global, Africa, Asia, Australia, Europe, N. America and Latin America. The section has extensive taxonomic coverage. In addition to the wide range of information for projects from various countries (e.g. Flora of China Project, Flora Malesiana Project) the major information of general and global scope lists the following:

   (i) **Authors.** Authors can be searched via gopher at Missouri Botanical Garden, St. Louis, USA.

   (ii) **Chromosome numbers.** Search maintained by Missouri Botanical Garden.

   (iii) **Collectors.** Botanical Collectors' Names from *Index Herbariorum*, Harvard University Herbarium, USA.

   (iv) **Glossaries** (Botanical Glossaries). This document gives access to glossaries of botanical terms on the internet. The Flora Australia glossary is provided by the Flora Section of the Australian Biological Resources Study (ABRS). The Flora of New South Wales glossary is provided by the National Herbarium of New South Wales (Herbarium, NSW).

   (v) **Names.** Garden Gate: Roots of Botanical Names.

   (vi) **BABEL.** A Multi-Lingual database for the Vernacular Names of European Wild Plants. Common names of European wild plants in all languages, a first attempt to contact those interested, and to assemble glossaries, dictionaries, databases.

   (vii) **The Gray Herbarium Index Database.** Gray Herbarium Index of New World Plants from the Harvard University Herbaria, Cambridge, Massachusetts, USA. The Gray Herbarium Index Database currently includes 287,225 records of New World vascular plant taxa at the level of species and below. The information is now accessible over the Internet via keyword searches from the E-mail Data Server and through the Biodiversity and Biological Collections Gopher.

   (viii) **SEPASAL.** Survey of Economic Plants for Arid and Semiarid Land is a major economic botany database on useful plants of drylands developed and maintained at the Royal Botanic Gardens, Kew, UK.

(ix) **TreeBASE Database.** TreeBASE is a pilot project of Harvard University Herbaria, USA sponsored by NSF, to establish a relational database of phylogenetic information.

(x) **Databases at Royal Botanic Gardens.** This powerful database at Kew, UK offers: **a)** List of the Vascular Plant Families and Genera. The data presented in this database are taken from the publication *Vascular Plant Families and Genera* compiled by R.K. Brummitt and published by the Royal Botanic Gardens, Kew in 1992; **b)** List of authors of plant names, based in *Authors of Plant Names*, ed. R.K. Brummitt and C.E. Powell and published by the Royal Botanic Gardens, Kew in 1992; and **c)** DNA Databases.

(xi) **GRIN Taxonomy Home Page.** Taxonomical databases of the Germplasm Resources Information Network of USDA's National Plant Germplasm System (NPGS) maintains record of economic plants with family and generic names nomenclature of the PEAS database, nomenclature of Seed Associations, noxious weeds, and rare and endangered plants.

(xii) **Index Nominum Genericorum (ING).** This is a collaborative project of the International Association for Plant Taxonomy (IAPT) and the Smithsonian Institution. The ING database covers validly published generic names of plants including fungi. In addition, ING includes bibliographic citations and information about the typification and nomenclatural status of generic names. Largely based on the Farr et al., (1979) and the supplement (Farr et al., 1986), the database is updated up to 1990, and even beyond. Generic names from the ING list that are recognised in the publication *Names in Current Use for Extant Plant Genera* (Greuter et al., 1993) are denoted with a [C] preceding the entry.

(xiii) **Indices Nominum Supragenericorum Plantarum Vascularium.** The database is maintained by the University of Maryland, College Park, Department of Plant Biology, USA. The project is a joint effort between the International Association for Plant Taxonomy and the Norton-Brown Herbarium at the University of Maryland. Its purpose is to capture all valid and legitimate extant vascular plants names proposed. above the rank of genus. These data are dynamic and constantly being updated so that any name listed is only the earliest valid and legitimate name found as of the latest update 22 March 1996.

(xiv) **The Integrated Taxonomic Information System (ITIS).** The database offers quality taxonomic information of flora and fauna of both aquatic and terrestrial habitats. ITIS is the product of a partnership of Federal agencies collaborating with systematists in the

Federal, state, and private sectors to provide taxonomic information. Geographic coverage will initially emphasise North American taxa.

(xv) **IOPI Database of Plant Databases (DPD).** This is a global list of databases from Australia to provide information about people who are compiling particular data. The DPD contains virtually no plant data as such, but serves to put you in contact with the projects that do. Most, not all, entries concern databases about higher plants.

(xvi) **Species Plantarum Project.** The Species Plantarum Project of International Organisation for Plant Information (IOPI) is a co-operative international project designed to help humanity manage the Earth's biodiversity efficiently and sustainably. It is a long-term project aiming to record essential taxonomic information on vascular plants on a world basis. It may be likened to a World Flora. It is expected that it will include accepted names and synonyms with places of publication and types, short descriptions of all taxa from family to infraspecific rank, keys, distributions, references to literary comments etc. It will be linked to the **Global Plant Checklist,** which is IOPI's first priority. The Checklist will form part of the **Species 2000** coverage of all organisms. IOPI's broad strategy is to merge data sources held by members (e.g. databases, floras, monographic treatments) into a computer-accessible format. The merged data will be edited for consistency by a worldwide network of specialists. The Checklist will become increasingly useful through refinements from taxonomic editing and the incorporation of additional datasets. It will be available as both a dynamic database and periodic hard copy publication.

(xvii) **The Global Plant Checklist Project (IOPI).** The project is organised by the IOPI (International Organisation for Plant Information) Checklist Committee. It will encompass about 300,000 vascular plant species and over 1,000,000 names; it is IOPI's first priority. Eventually, the Checklist will also include non-vascular plants (mosses, lichens, algae and liverworts). Development of the Global Plant Checklist Network is being supported by a grant 1997-98 from the Committee on Data for Science and Technology (CODATA) of the International Council of Scientific Unions (ICSU): A provisional checklist is in operation (186,664 records in early May 1997). IOPI also has a Species Plantarum Project (SPP) from Australia.

(xviii) **Names in Current Use (NCU).** Electronic version (NCU-3-e) of names in current use for Extant Plant Genera. The database has been generated from word processor files used for the camera-ready copy of the printed publication of NCU-3 edited by W. Greuter, B. Zimmer and W. Berendsohn published in 1993 by Koeltz

Scientific Books for the International Association. Numerous annotations and corrections have already been made so that in many parts it represents an update of the printed version. The user can query the database for plant genera and get the publication citation (place and date), the type species of the genus, nomenclatural details for vascular plants, algae, fungi and bryophytes.

(xix) **Monitoring Centre (UK).** The database provides information on nomenclatural, distribution and conservation status information on 82,500 species, subspecies and varieties of vascular plants worldwide (over 1/4 of the described higher plants). This taxon-based information is linked to 145,000 distribution records and 17,000 data source records.

(xx) **Species 2000, UK.** Species 2000 Program was established by the International Union of Biological Sciences (IUBS), in co-operation with the Committee on Data for Science and Technology (CODATA) and the International Union of Microbiological Societies (IUMS) in September 1994. It was subsequently endorsed by the UNEP Biodiversity Work Program 1996-1997 and associated with the Clearing House Mechanism of the UN Convention on Biological Diversity. Species 2000 has the objective of enumerating all known species of plants, animals, fungi and microbes on Earth as the baseline dataset for studies of global biodiversity. Users worldwide will be able to verify the scientific name, status and classification of any known species via the Species Locator on **Species 2000 Home Page**. The Species Locator provides access to species checklist data drawn from an array of participating databases. The user chooses between the stabilised Annual Checklist, and the Dynamic Checklist obtained by polling the databases on-line. It is intended to 1) operate a dynamic Common Access System on the Internet through which users can locate a species by name across an array of on-line taxonomic databases; 2) produce a stable species index, the Species 2000 Annual Checklist, available on the Internet and on CD-ROM, to be updated once a year; and 3) stimulate completion of the array of taxonomic databases by seeking resources both for the completion of existing databases, and to help establish new databases in order to cover identified gaps. The service will be made available as part of the Clearing House Mechanism under the United Nations Convention on Biological Diversity. Species 2000 will be operated by a federation of database organisations working closely with users, taxonomists and sponsoring agencies. The project is developing an indexing system for all groups of organisms, with an ultimate goal of listing all known species on Earth. From this simple Home Page one will be able to access Global Master Species Databases.

(xxi) **TRITON, Index to Organism Names.** The Index to Organism Names will represent the web public access portion of TRITON (Taxonomy Resource and Index To Organism Names). The name index will access data on all organism groups and currently holds data on fungi provided by the International Mycological Institute, on mosses provided by the Missouri Botanical Garden and on animals from the Zoological Record. Ongoing development work includes quality control, improved search options and incorporation of data for other organism groups and vernacular names. The Index will allow searching on taxon name and return basic nomenclatural and classification hierarchy data, with hypertext links to additional data.

(xxii) **TROPICOS Database.** The database is maintained by the Missouri Botanical Garden. This new access vehicle provides a real-time look at the data of over 750,000 scientific plant names. The records frequently have links to other associated names, types, synonymy, and bibliographic references.

8. **Conservation and Threatened Plants**. A number of projects of a general nature provide conservation-related information. **Biodiversity in Boreal Forests** is a joint project between the University of Helsinki, Finnish Environment Institute, Finnish Forest Industries Federation, and Finnish Forest Research Institute. The primary aim of this joint project is to increase our understanding of the ecology and dynamics of forest-dwelling species in boreal forests, and thereby introduce the presently lacking element of population biology and ecology into landscape ecological forest planning. The Convention on International Trade in Endangered Species of Wild Fauna and Flora (**CITES**), from UNEP (United Nations Environment Programme, Geneva, Switzerland) has established worldwide controls on the international trade in threatened species of animals and plants. On the website the user can find the CITES-listed species which occur in a certain country or find out when particular species were listed. Alternatively he can access the Resolutions and Decisions from past Conferences of the Parties and many other documents. Information is given in each of the three official languages used by CITES (English, French and Spanish). The mission of **Fauna and Flora International** (FFI) is to safeguard the future of endangered species of animals and plants. FFI, founded in 1903, is the world's longest established international mission. **The World Conservation Monitoring Centre** (WCMC), UK is also accessible via WCMC ftp. Here the viewer can find the Biodiversity Profile for India and Vietnam, Conservation Status Listing of Indian Plants, Plant Species Database (nomenclatural, distribution and conservation status information on 82,500 species, subspecies and varieties of vascular plants worldwide (over 1/4 of the described higher plants). This taxon-based

information is linked to 145,000 distribution records and 17,000 data source records. WCMC Library Catalogue is a searchable database of thousands of publications received by WCMC over the years. Threatened Plants of the World: Europe is the first-ever world list of threatened plants, compiled by WCMC in collaboration with IUCN-SSC (Species Survival Commission), the Royal Botanic Garden, Edinburgh and other leading botanical institutes. The data presented here are for Europe only.

9. **Economic Botany, Ethnobotany**. This link provides information on useful plants (food, medicine, material for textiles etc.) and harmful plants (plant pathology, poisonous plants, weeds). There are not many horticultural links (there are some in a separate file, Gardening). Only a few links on forestry and agriculture are provided (there are comprehensive lists on these topics, some of them under 'link collections, resource guides'). Some of the important sites are: Agricultural Network Information Center (**AgNIC**) Home Page, USA. AgNIC is a distributed network that provides access to agriculture-related information, subject area experts, and other resources in a manner in which the physical location of the resources is transparent to users. **AgNRIS** (Agricultural and Natural Resources Information Systems) Home Page is maintained in the Virginia Polytechnic Institute and State University, USA. **Forage Information System** provides links to worldwide forage-related information. Co-ordination of this effort is provided at Oregon State University, USA. **GrainGenes**, USA is a compilation of molecular and phenotypic information on wheat, barley, oats, rye and sugarcane. **Plant Resources of South-East Asia** (PROSEA), the Netherlands, is an international program focused on the documentation of information on plant resources of South-East Asia, covering the fields of agriculture, forestry, horticulture and botany. Useful information is provided in the Home Page.

10. **Gardening**. This link on gardening does not include Horticultural Societies, which are listed on a separate page, as are arboreta and botanical gardens, and news groups related to garden plants. The page has entries of general garden interest.

11. **Images**. Important information is provided on sources for images of plants, many of which can be downloaded. Agronomy & Horticulture 100 (**AgHrt 100**) Plant Images, Department of Agronomy & Horticulture, B.Y.U. Provo, UT, is an image gallery of JPEG images of approximately 150 economically important plants. Anyone is welcome to access this site and download images for non-commercial uses. **Andrew N. Gagg's Photo Flora** includes Subject lists of wild plant photos taken throughout Europe, following Flora Europaea, and is a photo library for writers, publishers etc. The work of Andrew N. Gagg, based in Worcester, UK, is a specialist collection, ever expanding, and covers

sites from many areas of Europe. **Botanical Scientific Illustrations**: an electronic herbarium site maintained by Victoria Vancek, and offering a sample of scientific botanical illustrations depicting flora from the Pacific west coast of British Columbia, Canada. **Flora of Europe**: a photographic herbarium, the Netherlands. 'Flora On-line' is meant to be an amateur photo-herbarium and when it started in March 1995 centered around (c.350) pictures of flowers, mostly of southern Europe. Until now it has been the work of one person, but anyone is invited to send in pictures (with descriptions) and join the Flora. **Grass Images** (Texas A&M University Bioinformatics Working Group). Taxa represented in the TAMU-BWG Image Database. Image files available here are mostly derived from collaboration with the Hunt Institute for Botanical Documentation and the Texas A&M University Press. **Photographic Collection of the Australian National Botanic Gardens** covering over 1 million acres of the Outer Coastal Plain in southern and central New Jersey. This website contains a collection of colour photographs of 45 pineland plants. Below each photograph are the scientific and common names of the plant, the family to which it belongs, its approximate height at maturity, a brief description of the plant and its habitat, and the place where the photograph was taken. **Plants Photo Gallery, USDA**, USA is sample of plant images being integrated into the PLANTS database for North American plants. These images include photographs of plants and plant habitats, with vernacular and scientific names, family name, photographer and location. **Smithsonian Catalog of Botanical Illustrations** is a database of the Department of Botany of the Smithsonian Institution's National Museum of Natural History, USA, with more than 3000 botanical illustrations curated by the department's scientific illustrator, Alice Tangerini. As part of a long-term project to make an on-line illustrated catalog of these illustrations available for their staff and others needing access to this information, the department offers 500 images in the three families that have been completed: Bromeliaceae, Cactaceae, and Melastomataceae.

12. **Journals, Books, Literature Databases, Publishers**. The page provides detailed listings and links, many of which have already been included on the main page of IDB.

13. **Link Collections, Resource Guides**. Information on important links which can be very useful for searching information on the web. **PlantLink** by PlantAmerica is a search engine for botanists. In an effort to find better ways to harness the wealth of information on the Worldwide Web, PlantAmerica has introduced a series of customised search engines and filtering tools called 'The Learning Network'. The backbone of 'The Learning Network' is 'PlantAmerica's DataNet', a

customised version of the 85,000 entry USDA plant list—to which a cluster of tools and educational resources are being added. The latest addition, 'PlantAmerica's CustomSearch' was developed under the direction of Dr. Eric Marler, formerly of IBM, recently the cofounder of the Global Health Network, and now a passionate horticulturalist and botanist. The resource is a combination of: 1) listing of plants—by common, family and genus/species names; 2) a sorting filter of plant characteristics—bud, bark, seed, root, propagation, disease etc., 3) a second filter that sorts alphabetically by keyword or domain; 4) the query is assigned a unique code which is automatically attached to the search request; and 5) The search engine in turn simultaneously uses 10 of the most powerful search tools on the Internet to locate and form links to the URLs that specifically meet the chosen criteria. The user can interface with 'PlantAmerica's CustomSearch' directly at www.plantamerica.com. However, by virtue of technology recently incorporated, PlantAmerica will soon be able to supply URLs so that sponsors can also 'tag' plant names or plant images at their website. The user, by simply clicking on the 'tagged' item, can link directly with the specific search query. **The Worldwide Web Virtual Library: Biosciences** is a massive index to the bioscience resources of the Net, maintained by Keith Robison of the Harvard University Biological Laboratories, USA. Subjects covered include biological molecules, biotechnology, genetics, immunology, plant biology, and many, many more. URLs can be quickly searched using the search form for Biosciences: WWW Virtual Library: Botany-Plant Biology (University of Oklahoma, USA), maintained by Scott Russell. **Ag-Links**, Worldwide Websites of Interest to Agriculture, from the Gennis Agency, USA. **Biodiversity and Biological Collections Web Server Botany page** is a WWW server devoted to information of interest to systematists and other biologists of the organismic kind. Within these pages the user will find information about specimens in biological collections, taxonomic authority files, directories of biologists, reports by various standard bodies (IOPI, ASC, SA2000 etc.), an archive of the Taxacom and MUSE-L listserves, access to on-line journals (including Flora On-line) and information about biodiversity and collection oriented projects (MUSE and NEODAT). Recently added are index images to the Biological Image Archive BBCWS Gopher. **BIOSIS**, Biosystematics and Life Science Resources, UK provides access to information of interest to all biologists, but with particular emphasis on the fields of taxonomy and nomenclature. Users will find access to primary services for biologists, such as organism name and nomenclatural related services being developed in conjunction with production of the Zoological Record. **Plant Science Links** maintained by the University of Reading provides direct links to several important sites on the Internet including: Chromosome Numbers (search), The Cyber-Plantsman, Flora of North America, The Garden Gate Garden Net, International Organisation of Palaeobotany, International Organization for Plant Information, Lichen Herbarium Swedish Museum of Natural History, The Linne Herbarium, Swedish Museum of Natural History, British Trees, List of WWW Sites of Interest to Ecologists, Missouri Botanical Garden, Natural History Book Service—Botany Books, The Natural History Museum (London), Palynology and Paleoclimatology, Plant Genome Server, Plant Taxonomists Online, The Royal Botanic Gardens Kew, TAXACOM List Serve Archives, List of WWW Sites of Interest to Ecologists, The Tree of Life Home Page, TROPICOS via Remote Managing Gigabytes, UK Plant Genetic Resources Group, University of California, Berkeley Museum (Public Exhibits),Vatican Exhibit— Herbals, Garden Web (including wild flowers), The Institute for Plant Diseases and Plant Protection (IPP), Department Horticulture, Hannover, Germany, and Department Horticulture, Osnabrueck, Germany.

14. **Listserves and NewsGroups**. Lists of taxonomic interest include CACTUS-L and TAXACOM. Important news groups include BIOSCI and Garden Plants.

15. **Lower Plants**. The page provides information on algae, bryophytes and fungi. The general source for all these groups is **Botany 3100**, Survey of Non-Vascular Plants, California State University, Stanislaus, USA, Fall 1996. In addition there are numerous group-wise listings with links.

16. **Other Resources**. This page contains Links that do not fit well into the other current categories, or links waiting to be moved elsewhere: Plant Genetics, Seeds, Standards and Others. The page lists **Draft BioCode**, the prospective international rules for the scientific names of organisms, prepared and edited by W. Greuter et al. (the IUBS/IUMS International Committee for Bionomenclature).

17. **Palaeobotany, Palynology, Pollen**. Among the prominent listings are the American Association of Stratigraphic Palynologists (AASP) from the University of Toronto, Canada, International Federation of Palynological Societies (IFPS), International Organization of Paleobotany etc. Northwest European Pollen Flora, a vegetation classification and identification manual of pollen and spores in the British Museum of Natural History, UK is a project which aims to publish keys and detailed descriptions of the pollen grains of all seed plant families and certain important fern and moss families occurring in north-west Europe.

18. **Software**.The page lists a large number of softwares useful for botanical activities. The major ones include **Aditsite** (Microsoft Windows software for environmental assessment, wildlife recording and species identification), **IOPAK** (Software for Computing Plant Biomass) and

BIOSIS—Software Reviews, UK. Links to over 150 reviews of biological software packages are given. Each review contains the following elements: software title, including version, information; author(s)' and publisher's name and contact information; description of the manual if included; publication year; an abstract containing specifications for running the software. **COMPARE**, from the University of Oregon, USA presents computer programs for the phylogenetic analysis of comparative data. COMPARE consists of several programs which analyse comparative or interspecific data while taking phylogenetic information into account. **DELTA** Australia (Descriptive Language for Taxonomy) is a Worldwide Web service for the DELTA format and suite of programs devised by Mike Dallwitz of the CSIRO Division of Entomology in Canberra, Australia. It also includes shells (DIANA, DELTA MENU SYSTEM, TAXASOFT) to run the DELTA programs, and datasets for INTKEY. The **Digital Taxonomy** Website by Mauro J. Cavalcanti is an attempt to present a wide-ranging resource of information for biodiversity data management in the Worldwide Web, and promote the effective use of computers for handling biological software development projects. Digital Taxonomy provides a range of links on software, hardware, methodologies, standards, data sources, and projects related to biodiversity data management, with emphasis on the exchange of free scientific software tools, computer techniques, and Internet addresses with links to other sources of information. The **PANDORA** taxonomic database system, created by Richard J. Pankhurst, UK, is a database system for biodiversity research projects, such as floras or monographs, and is the official database used at the Royal Botanic Garden, Edinburgh (RBGE) for taxonomic datasets. PANDORA can also be used for maintaining catalogues of collections such as herbarium specimens and a herbarium label printing system is included. A completely functional demonstration version of PANDORA is available by FTP, or via standard mail on three diskettes. **Platypus**, a Database Package for Taxonomists, is a Microsoft Windows-based program for managing taxonomic, geographic, ecological and bibliographic information. Platypus may be trialled free of charge by downloading a demo version. **TRANSLAT**, Botanical Latin Translation Program is a free-ware program by Peter Bostock, Queensland Herbarium, Australia (available also from the Missouri Botanical Garden FTP server, St. Louis, USA). **WINLABEL** by Lanius Software (Herbarium Label-making Software) is a commercial, relational database program designed to maintain records of plants the viewer has collected for placement in an herbarium. Its primary purpose is to maintain detailed records of collected plant specimens in a format that permits the user to generate one or more herbarium labels for each collected specimen. WINLABEL features an intuitive, graphical user interface and swift data entry. Printed labels by county, state, location, etc. are available. Labels are formatted to report all the user's data in a clean, consistent and professional manner; supports database entry for any state, county, or province in North America. Customisable header and footer text; options for printing graphics. **XID Expert System**, referring to Identification, Prescription and Description Database Software, USA, is a relational database for identification.

19. **Vascular plant Families**. The page provides information on Pteridophytes, Gymnosperms and Angiosperms. Coverage is region-wise for Africa, Asia, Australia and New Zealand, Europe, Latin America and N America (Canada, USA). Angiosperms have a very broad coverage with alphabetic listing of a very large number of families. In addition the main databases include:

   (i) **Concordance of Family Names**. This database was prepared by James L. Reveal, Department of Plant Biology, University of Maryland, USA, at the request of the U.S. Department of Agriculture's Animal and Plant Health Inspection Service and is intended as a quick search database for family names and their alternative uses by Cronquist, Dahlgren and Thorne. When available, Takhtajan's nomenclature will be added. Fully annotated phylogenetic arrangements of the flowering plants are also available from the same source for Cronquist, Dahlgren and Thorne.

   (ii) **Flowering Plant Gateway**. Developed by Texas A & M University Bioinformatics Group, the Gateway offers various paths for exploration or comparison of two major systems of flowering plant classification. Family-level data include links to WWW information for those families for which information is available.

   (iii) **Genus of the Week**. This site is maintained by Jennifer Forman, Biology Department, University of Massachusetts, Boston, USA. The site features a different genus each week, giving information about each plant, its growth habit, etymology of the plant name, interesting notes, and selected links to images and references elsewhere.

   (iv) **Plant Family Names**. Developed by James L. Reveal, Department of Plant Biology, University of Maryland, USA, the data include Extant Vascular Plant Family Names, Extant Vascular Plant Family Names in Current Use, Extant Vascular Plant Ordinal Names, and Extant Vascular Plant Superordinal Names. The information includes name and synonymy with authorship and dates of publication. As more information is obtained of the ranks of division/phylum and class, these will be added.

References

Adanson, M. (1763). *Familles des plantes*, Paris, 2 vols.

Alston, R.E. and B.L. Turner. (1963). Natural hybridisation among four species of *Baptisia* (Leguminosae). *Amer. J. Bot.* **50**: 159-173.

Arber, A (1938). *Herbals: Their Origin and Evolution* (2nd ed.). Cambridge Univ. Press.

Arber, E.A.N. and J. Parkins. (1907). On the Origin of Angiosperms. *Bot. J. Linn. Soc.* **38**: 29-80.

Ashlock, P.H. (1971). Monophyly and related terms. *Syst. Zool.* **20**: 63-69.

Axelrod, D.I. (1970). Mesozoic paleogeography and early angiosperm history. Bot. Rev. **36**: 277-319.

Babcock, E.B. (1947). The genus *Crepis* pt. 1. The taxonomy, phylogeny distribution and evolution of *Crepis*. *Univ. Calif. Publs. Bot.* **21**: 1-197.

Bailey, I.W. (1944). The development of vessels in angiosperms and its significance in morphological research. *Am. J. Bot.* **31**: 421-428.

Bailey, L.H. (1949). *Manual of Cultivated Plants* (rev. Ed.). Macmillan, New York.

Barber, H.N. (1970). Hybridization and evolution of plants. *Taxon* **19**: 154-160.

Barthlott, W. (1981). Epidermal and seed surface characters of plants: Systematic applicability and some evolutionary Aspects. *Nordic J. Bot.* **1**: 345-355.

Barthlott, W. and D. Froelich. (1983). Mikromorphologie und Orientierungsmuster epicuticularer Wachs-Kristalloide: Ein neues systematisches Merkmal bei Monocotylen. *Pl. Syst. Evol.* **142**: 171-185.

Barthlott, W. and G. Voit. (1979). Mikromorphologie der Samenschalen und Taxonomie der Cactaceae Ein raster-elektronem-microscopischer uberblick. *Plant Syst. Evol.* **132**: 205-229.

Bate-Smith, E.C. (1958). Plant phenolics as taxonomic guides. *Proc. Linn. Soc. Lond.* **169**: 198-211.

Bate-Smith, E.C. (1962). The phenolic constituents of plants and their taxonomic significance. *J. Linn. Soc. Bot.* **58**: 95-173.

Bate-Smith, E.C. (1968). The phenolic constituents of plants and their taxonomic significance. *J. Linn. Soc. Bot.* **60**: 325-383.

Bauhin, C. (1596). *Phytopinax seu enumeratic plantarum.....* Basel.

Bauhin, C. (1623). *Pinax theatri botanici.* Basel.

Baum, B.R. (1977). *Oats: Wild and Cultivated. A Monograph of the Genus Avena* L. (*Poaceae*). Minister of Supply and Services, Ottawa.

Behnke, H.D. (1965). Über das phloem der Dioscoreaceen unter besonderer berücksichtigung ihrer phloembecken. II. Mitteilung: Elektronenoptische untersuchungen zur feinstruktur des phloembeckens. *Z. Pflanzenphysiol.* **53**: 214-244.

Behnke, H.D. (1976). Ultrastructure of sieve-element plastids in Caryophyllales (Centrospermae); evidence for the delimitation and classification of the order. *Plant Syst. Evol.* **126**: 31-54.

Behnke, H.D. (1977). Transmission electron microscopy and systematics of flowering plants. In: K. Kubitzki. *Flowering Plants Evolution and Classification of Higher Categories. Plant Syst. Evol. Suppl. 1*, 155-178.

Behnke, H.D. (1997). Sarcobataceae a new family of Caryophyllales. *Taxon* **46**: 495-507.

Behnke, H.D. and W. Barthlott (1983). New evidence from the ultrastructural and micromorphological fields in Angiosperm classification. *Nordic. J. Bot.* **1**: 341-460.

Belfod, H.S. and W.F. Thomson (1979). Single copy DNA homologies and phylogeny of *Atriplex*. Carnegie Inst. Wash Year Book **78**: 217-223.

Bell, G.A. (1971). Comparative Biochemistry of non-protein amino acids. In: J.B. Harborne, D. Boulter and B.L. Turner (eds.) *Chemotaxonomy of Leguminosae*. Academic Press, London, pp. 179-206.

Bentham, G. (1858). *Handbook of British Flora*. (7ᵗʰ ed., revised by A.B. Rendle in 1930). Ashford, Kent.

Bentham, G. (1863-1878). *Flora Australiensis.*London, 7 volumes.

Bentham, G. and J.D. Hooker. (1862-83). *Genera Plantarum*. London, 3 vols.

Bessey, C.E. (1915). Phylogenetic taxonomy of flowering plants. *Ann. Mo. Bot. Gard.* **2**: 109-164.

Bhandari, M.M. (1978). *Flora of the Indian Desert*. Sc. Publ., Jodhpur.

Blackith, R.E. and R.A. Reyment. (1971). *Multivariate Morphometrics*. Acad. Press, London.

Blakeslee, A.F. , A.G. Avery, S. Satina and J. Rietsama. (1959). *The genus Datura*. Ronald Press, New York.

Bonnett, H.T. and E.H. Newcomb. (1965). Polyribosomes and cisternal accumulations in root cells of radish. *J. Cell Biol.* **27**: 423-432.

Bordet, J. (1899). Sur l'agglutination et la dissolution des globules rouges par le serum d'animaux injectes de sang defibriné. *Ann. Inst. Pasteur* **13**: 225-250.

Boulter, D. (1974). The use of amino acid sequence data in the classification of higher plants. In: G. Bendz and J. Santesson (eds.). *Chemistry and Botanical Classification, Nobel Symposium*, **25**, pp. 211-216. Acad. Press, London/New York.

Brenner, G.H. and I. Bickoff. (1992). Palynology and age of the Lower Cretaceous basal Kurnub Group from the coastal plain to the northern Negev of Israel. *Palynology* **16**: 137-185.

Brenner, G.J. (1963). Spores and Pollen of Potomac Group of Maryland. *Maryland Department of Geology, Mines, Water Resources Bulletin* **27**: 1-215.

Brenner, G.L. (1996). Evidence for the earliest stage of angiosperm pollen evolution. A paleoequatorial section from Israel. In: D.W. Taylor and L.J. Hickey (eds.) *Flowering Plant Origin, Evolution and Phylogeny*. Chapman & Hall Inc., New York, pp. 91-115.

Brown, R.W. (1956). Palmlike plants from the Delores Formation (Triassic) in southwestern Colorado. *U. S. Geological Survey Professional Paper* **274**: 205-209.

Brunfels, O. (1530). *Herbarium vivae eicones*. Argentorati, 3 tomes.

Caesalpino, A. (1583). *De plantis libri*. Florentiae.

Cain, A.J. and G.A. Harrison. (1958). An analysis of the taxonomists's judgement of affinity. *Proc. Zool. Soc. Lond.* **131**: 85-98.

Cain, A.J. and G.A. Harrison. (1960). Phyletic weighting. *Proc. Zool. Soc. Lond.* **135**: 1-31.

Candolle, A.P. de (1813). *Theorie elementaire de la botanique*. Paris.

Candolle, A.P. de (1824-73). *Prodromus systematis naturalis regni vegetabilis*. Paris, 17 vols.

Carlquist, S. (1987). Presence of vessels in *Sarcanda* (Chloranthaceae); comments on vessel origin in angiosperms. *Amer. J. Bot.* **64**: 1765-1771.

Carlquist, S. (1996). Wood anatomy of primitive Angiosperms: New perspective and syntheses. In: D.W. Taylor and L.J. Hickey (eds.) *Flowering Plant Origin, Evolution and Phylogeny*. Chapman & Hall Inc., New York, pp. 68-90.

Chase, M.W., D.E. Soltis, R.G. Olmstead et al. (1993). Phylogenetics of seed plants: an analysis of nucleotide sequences from the plastid gene rbcL. *Ann. Missourie. Bot. Gdn.* **80**: 528-580.

Cheadle, V.I. (1953). Independent origin of vessels in the monocotyledons and dicotyledons. *Phytomorphology* **3**: 23-44.

Clark, P.J. (1952). An extension of the coefficient of divergence for use with multiple characters. *Copeia* **2**: 61-64.

Clifford, H.T. (1977). Quantitative studies of inter-relationships amongst the Liliatae. In: K. Kubitzki. *Flowering Plants Evolution and Classification of Higher Categories. Pl. Syst. Evol. Suppl.* **1**, 77-95.

Colless, D.H. (1967). An examination of certain concepts in phenetic taxonomy. *Syst. Zool.* **16**: 6-27.

Collett, H. (1921). *Flora Simlensis* 2nd ed. Thacker, Spink, Calcutta.

Coode, M.J.E. (1967). Revision of Genus *Valerianella* in Turkey. *Notes Roy. Bot. Gard. Edinb.* **27**: 219-256.

Cornet, B. (1986). Reproductive structures and leaf venation of Late Triassic angiosperm, *Sanmiguelia lewisii. Evol. Theory* **7**:231-309.

Cornet, B. (1989). Reproductive morphology and biology of *Sanmiguelia lewisii* and its bearing on angiosperm evolution in Late Triassic. *Evol. Trends Plants* **3**: 25-51.

Cornet, B. (1993). Dicot-like leaf and flowers from the Late Triassic tropical Newark Supergroup rift zone, U. S. A. *Modern Geol.* **19**: 81-99.

Cornet, B. (1996). A New Gnetophyte from the Late Carnian (Late Triassic) of Texas and its bearing on the origin of the angiosperm carpel and stamen. In: D.W. Taylor and L.J. Hickey (eds.) *Flowering Plant Origin, Evolution and Phylogeny.* Chapman & Hall Inc., New York, pp. 32-67.

Couper, R.A. (1958). British Mesozoic microspores and pollen grains. *Palaeontographica*, Abt. B, **103**: 75-179.

Crane, P.R., E.M. Friis and K.R. Pedersen. (1995). The origin and early diversification of angiosperms. *Nature* **374**: 27-33.

Crawford, D.J. and E.A. Julian. (1976). Seed protein profiles in the narrow-leaved species of *Chenopodium* of the Western United States: Taxonomic value and comparison with distribution of flavonoid compounds. *Am. J. Bot.* **63**: 302-308.

Cronquist, A. (1968). *The Evolution and Classification of Flowering Plants.* Houghton Mifflin, New York.

Cronquist, A. (1977). On the taxonomic significance of secondary metabolites in Angiosperms. *Plant Syst. Evol.,* Suppl. **1**: 179-189.

Cronquist, A. (1981). *An Integrated System of Classification of Angiosperms.* Columbia Univ. Press, New York.

Cronquist, A. (1988). *The Evolution and Classification of Flowering Plants.* (2nd ed.). New York Botanical Garden, Bronx, New York.

Cronquist, A., A.L. Takhtajan and W. Zimmerman (1966). On the higher taxa of Embryobionta. *Taxon* **15**: 129-134.

Dahlgren, G. (1989). An updated angiosperm classification. *Bot. J. Linn. Soc.* **100**: 197-203.

Dahlgren, G. (1989). The last Dahlgrenogram. System of classification of dicotyledons, pp. 249-260. In: K. Tan (ed.). *The Davis and Hedge Festschrift.* Edinburgh Univ. Press, Edinburgh, pp. 249-260.

Dahlgren, R.M.T. (1975). A system of classification of angiosperms to be used to demonstrate the distribution of characters. *Bot. Notiser* **128**: 119-147.

Dahlgren, R.M.T. (1977). Commentary on a Diagrammatic Presentation of the Angiosperms. In: K. Kubitzki (ed.). *Flowering Plants: Evolution and Classification of Higher Categories.* Plant Systematics and Evolution Suppl. 1. Springer-Verlag Wien/New York, pp. 253-283.

Dahlgren, R.M.T. (1980). A revised system of classification of angiosperms. *Bot. J. Linn. Soc.* **80**: 91-124.

Dahlgren, R.M.T. (1983). General aspects of angiosperm evolution and macrosystematics. *Nordic. J. Bot.* **3**: 119-149

Dahlgren, R.M.T., H.T. Clifford and P.F. Yeo. (1985). *The Families of Monocotyledons.* Springer-Verlag, Berlin.

Dahlgren, R.M.T. and F.N. Rasmussen. (1983). Monocotyledon evolution: characters and phylogenetic estimation. *Evol. Biol.* **16**: 255-395.

Dahlgren, R.M.T., S. Rosendal-Jensen and B.J. Nielsen. (1981). A revised classification of the angiosperms with comments on the correlation between chemical and other characters, In: D.A. Young and D.S. Seigler (eds.). *Phytochemistry and Angiosperm Phylogeny.* Praeger, New York, pp. 149-199.

Darlington, C.D. and E.K. Janaki-Ammal. (1945). *Chromosome Atlas of Cultivated Plants.* Allen and Unwin, London.

Darlington, C.D. and A.P. Wylie. (1955). *Chromosome Atlas of Flowering Plants.* Allen and Unwin, London.

Darwin, C. (1859). *The Origin of Species*. London.

Daugherty, L.H. (1941). *The Upper Triassic Flora of Arizona with a Discussion on its Geological Occurrence*. Contributions to Paleontology 526, Carnegie Institution of Washington.

Davis, P.H. (1960). Materials for the Flora of Turkey. IV. Ranunculaceae, II. *Notes Roy. Bot. Gard. Edinburgh*, **23**: 103-161.

Davis, P.H. and V.H. Heywood. (1963). *Principles of Angiosperm Taxonomy*. Oliver and Boyd, London.

Donoghue, M.J. and J.A. Doyle. (1989). Phylogenetic analysis of angiosperms and the relationships Hamamelidae. In: P.R. Crane and S. Blackmore (eds.). *Evolution, Systematics and Fossil History of the Hamamelidae*. Clarendon Press, Oxford, pp. 17-45.

Doyle, J.A. (1969). Cretaceous angiosperm pollen of Atlantic Coastal Plain and its evolutionary significance. *J. Arnold Arboretum*. **50**: 1-35.

Doyle, J.A. (1978). Origin of angiosperms. *Ann. Rev. Ecol. Systematics*. **9**: 365-392.

Doyle, J.A., M. Van Campo and B. Lugardon. (1975). Observations on exine structure of Eucommiidites and Lower Cretaceous angiosperm pollen. *Pollen and Spores* **17**: 429-486.

Doyle, J.A. and M.J. Donoghue. (1987). The origin of angiosperms: a cladistic approach. In: E.M. Friis, W.G. Chaloner and P.R. Crane (eds.). *The Origins of Angiosperms and Their Biological Consequences*. Cambridge University Press, U. K., pp. 17-49.

Doyle, J.A. M.J. and M.J. Donoghue. (1993). Phylogenies and angiosperm diversification. *Paleobiology* **19**: 141-167.

Du Rietz. (1930). Fundamental units of biological taxonomy. *Svensk bot. Tidskr*. **24**: 333-428.

Dykes, W.R. (1913). *The Genus Iris*.

Eames, A.J. (1961). *Morphology of Angiosperms*. McGraw-Hill Book Co., New York.

Ehrendorfer, F. (1968). Geographical and ecological aspects of infraspecific diiferentiation . In: V.H. Heywood (ed.). *Modern Methods in Plant Taxonomy* Acad. Press, New York, pp. 261-296.

Ehrendorfer, F. (1983). Summary Statement. *Nord. J. Bot*. **3**: 151-155.

Eichler, A.W. (1883). *Syllabus der Vorlesungen über Specielle und Medicinisch-Pharmaceutische Botanik*. Leipzig.

Engler, A. (1892). *Syllabus der Pflanzenfamilien*. Berlin.

Engler, A. (H. Melchior and E. Werdermann, eds.). (1954). *Syllabus der pflanzenfamilien*.12th ed., vol. 1. Gebruder Borntraeger, Berlin.

Engler, A. (H. Melchior, ed.). (1964). *Syllabus der pflanzenfamilien*.12th ed., vol. 2. Gebruder Borntraeger, Berlin.

Engler, A. (ed.). (1900-1953). *Das Pflanzenreich*. Regni vegetabilis conspectus Im Auftrage der Preus. Akademie der Wissenschaften, Leransgegeben von A. Engler, Berlin. (after Engler's death subsequent volumes, continuing upto 1953 were edited by other authors)

Engler, A. and L. Diels (1936). *Syllabus der pflanzenfamilien*. 11th ed. Berlin.

Engler, A. and K. Prantl (1887-1915). *Die naturlichen pflanzenfamilien*. Leipzig, 23 vols.

Erdtman, G. (1948). Did dicotyledonous plants exist in Early Jurrassic time? *Geol. Fören. Stockholm Förh*. **70**: 265-271.

Erdtman, G. (1966). *Pollen Morphology and Plant Taxonomy. Angiosperms. (An Introduction to Palynology. I.)*. Hafner Publ. Co., London.

Fairbrothers, D.E. (1983). Evidence from nucleic acid and protein chemistry, in particular serology, in angiosperm classification. *Nordic. J. Bot*. **3**: 35-41.

Farr, E.R., J.A. Leussink and F.A. Stafleu (eds.). (1979). *Index Nominum Genericorum (Plantarum)*. *Regnum Veg*. **100-102**: 1-1896.

Farr, E.R., J.A. Leussink and G. Zijlstra. (eds.). (1986). *Index Nominum Genericorum (Plantarum) Supplementum* I. *Regnum Veg*. **113**: 1-126.

Fassett, N.C. (1957). *A Manual of Aquatic Plants*. Univ. Wisconin Press, Madison.

Faust, W.Z. and S.B. Jones. (1973). The systematic value of trichome complements in a North American Group of *Vernonia* (Compositae). *Rhodora* **75**: 517-528.

Federov, A.A. (ed.). (1969). *Chromosome Numbers of Flowering Plants*. Akad. Nauk SSSR, Leningrad.

Fiori, A. and G. Paoletti. (1896). Flora analitica d'Italia **1**: 1-256. Padova.

Friedrich, H.C. (1956). Studien über die natürliche verwandtschaft der Plumbaginales und Centrospermae. *Phyton (Austria)* **6**: 220-263.

Frodin, D.G. (1984). *Guide to the Standard Floras of the World.* Cambridge Univ. Press.

Frost, F.H. (1930). Specialization in secondary xylem in dicotyledons. I. Origin of vessels. *Botanical Gazette* **89**: 67-94.

Gagnepain, F. And Boureau. (1947). Nouvelles considerations systématische á propos du Sarcopus abberans Gagnepain. *Bull. Soc. Bot. Fr.* **94**: 182-185.

Garnock-Jones, P.J. and C.J. Webb. (1996). The requirement to cite authors of plant names in botanical journals. *Taxon*, **45**: 285-286.

Gaussen, H. (1946). *Les Gymnosperms actuelles et fossiles.* Pt. 3. Travaux du Laboratoire Forestier, Toulouse.

Gershenzon, J. and T.J. Mabry. (1983). Secondary metabolites and the higher classification of angiosperms. *Nordic J. Bot.* **3**: 5-34.

Gottsberger, G. (1974). Structure and function of primitive angiosperm flower A Discussion. *Acta Bot. Neerl.* **23**: 461-471.

Gower, J.C. (1966). Some distance properties of latent root and vector methods used in multivariate analysis. *Biometrika* **53**: 325-338.

Grant, V. (1957). The plant species in theory and practice. In: E. Mayr (ed.). *The Species Problem.* Amer. Assoc. Adv. Sci. Washington, D. C., pp. 39-80.

Grant, V. (1981). *Plant Speciation* (2nd ed.). Columbia Univ. Press, New York.

Gregory, W.C. (1941). Phylogenetic and cytological studies in the Ranunculaceae. *Trans. Am. Phil. Soc.* **31**: 443-520.

Greuter, W., R.K. Brummitt, E. Farr, N. Kilian, P.M.Kirk and P.C. Silva. (1993). *Names in current use for extant plant genera.* Koeltz, Königstein, Germany. xxvii + 1464 pp. *Regnum veg.* Vol. **129**.

Greuter, W. et al. (1994). International code of botanical nomenclature (Tokyo Code) adopted by the Fifteenth International Botanical Congress, Yokohama, August-September 1993. *Regnum Veg.* **131**.

Gunderson, A. (1939). Flower buds and phylogeny of dicotyledons. *Bull. Torrey Bot. Club* **66**: 287-295.

Hall, D.W. (1981). Microwave: a method to control herbarium insects. *Taxon*, **30**: 818-819.

Hanelt, P. and J. Schultze-Motel. (1983). Proposal (715) to conserve *Triticum aestivum* L. (1753) against *Triticum hybernum* L. (1753) (Gramineae). *Taxon* **32**: 492-498.

Harborne, J.B. and B.L. Turner. (1984). *Plant Chemosystematics.* Acad. Press, London.

Harris, T.M. (1932). The fossil flora of Scorseby Sound East Greenland. Part 3: Caytoniales and Bennettitales. *Meddelelser om Grônland* **85**(5): 1-133.

Hawksworth, D.L. 1995. Steps along the road to a harmonized bionomenclature. *Taxon*, **44**: 447-456.

Hedge, I.C. and J.M. Lamond. (1972). Umbelliferae. Multi-accesskey to the Turkish genera. In: P.H. Davis (ed.). *Flora of Turkey.* Edinburgh Univ. Press, Edinburgh, vol. 4, pp. 171-177.

Henderson, D.M. (1983). *International Directory of Botanical Gardens IV.* Koeltz, Koenigstein.

Hennig, W. (1950). *Grundzüge einer Theorie der phylogenetischen Systematik.* Deutscher Zentralverlag, Berlin.

Hennig, W. (1957). Systematik und Phylogenese. *Ber. Hundertjahrfeier Deutsch. Entomol. Ges.*, pp. 50-70.

Hennig, W. (1966). *Phylogenetic Systematics.* Translated by D.D. Davies and R. Zangerl. Univ. Illinois Press, Urbana.

Heslop-Harrison, J. (1952). A reconsideration of plant teratology. *Phyton* **4**: 19-34.

Heslop-Harrison, J. (1958). The unisexual flower a reply to criticism. *Phytomorphology* **8**: 177-184.

Hickey, L.J. and J.A. Doyle. (1977). Early Cretaceous fossil evidence for angiosperm evolution. *Bot. Rev.* **43**: 1-104

Hickey, L.J. and D.W. Taylor. (1992). Paleobiology of early angiosperms: evidence from sedimentological associations in Early Cretaceous Potomac Group of eastern USA *Paleontological Soc. Spec. Publ.* **6**: 128.

Hickey, L.J and D.W. Taylor. (1996). Origin of angiosperm flower. In: D.W. Taylor and L. J. Hickey (eds.) *Flowering Plant Origin, Evolution and Phylogeny.* Chapman & Hall Inc., New York, pp. 176-231.

Hill, S.R. (1983). Microwave and the herbarium specimens: potential dangers. *Taxon* **32**: 614-615.

Holmgren, P.K., N.H. Holmgren and L.C. Barnett. (1990). Index herbariorum. Part I: The Herbaria of the World (8th ed.). *Regnum Veg.* **120**.

Hooker, J.D. (1870). *Student's Flora of British Isles.*Macmillan and Co., London. (3rd ed., 1884).

Hooker, J.D. (1872-97). *Flora of British India..* L. Reeve and Co., London, 7 vols.

Hughes, N.F. (1961). Further interpretation of *Eucommiidites* Erdtman, 1948. *Palaeontology* **4**: 292-299.

Hutchinson, J. (1946). *A Botanist in South Africa.* London.

Hutchinson, J. (1948). *British Flowering Plants.* London.

Hutchinson, J. (1964-67). *The Genera of Flowering Plants.* Clarendon, Oxford, 2 vols.

Hutchinson, J. (1968). *Key to the Families of Flowering Plants of the World.* Clarendon, Oxford, 117 pp.

Hutchinson, J. (1969). *Evolution and Phylogeny of Flowering Plants.* Acad. Press, London.

Hutchinson, J. (1973). *The Families of Flowering Plants.* (3rd ed.). Oxford Univ. Press. (2nd 3d. 1959; Ist ed. 1926, 1934)

Hutchinson, J and J.M. Dalzeil. (1927-1929). *Flora of West Tropical Africa.* London, 2 vols.

Index Kewensis plantarum phanerogamarum. (1893-95), 2 vols. Oxford. 16 *supplements* up to 1971.

Jaccard, P. (1908). Nouvelles recherches sur la distribution florale. *Bull. Soc. Vaud. Sci. Nat.* **44**: 223-270.

Jardine, N. and R. Sibson. (1971). *Mathematical Taxonomy.* Wiley, London.

Johnson, B.L. (1972). Seed protein profiles and the origin of the hexaploid wheats. *Amer. J. Bot.* **59**: 952-960.

Jones, S.B. Jr. and A.E. Luchsinger. (1986). *Plant Systematics,* 2nd ed. McGraw-Hill Book Co., New York.

Jordan, A. (1873). Remarques sur le fait de l'existence en société à l'état sauvage des espéces végétales affines. *Bull. Ass. Fr. Avanc. Sci.* **2**, session Lyon.

Jussieu, A.L. de (1789). *Genera plantarum.* Paris.

Kerguélen, M. (1980). Proposal (68) on article 57.2 to correct the *Triticum* example. *Taxon* **29**: 516-517.

Kluge, A.G. and J.S. Farris. (1969). Quantitative phyletics and the evolution of anurans. *Syst. Zool.* **18**: 1-32.

Komarov, V.L. And B.K. Shishkin. (1934-1964). *Flora SSSR.* AN SSSR Press, Moscow/Leningrad, 30 vols.

Kosakai, H., M.F. Moseley and V.I. Cheadle. (1970). Morphological studies in the Nymphaeaceae. V. Does *Nelumbo* have vessels?. *Amer. J. Bot.* **57**: 487-494.

Krassilov, V.A. (1977). Contributions to the knowledge of the Caytoniales. *Rev. Paleobot. Palynology* **24**: 155-178.

Kraus, R. (1897). Über Specifishe Reactionin in Keimfreien Filtraten aus Cholera, Typhus und Pestbouillon Culturen erzeugt durch homologes Serum. *Weiner Klin. Wechenschr.* **10**: 136-138.

Kubitzki, K. (ed.) (1993). 'The Families and Genera of Vascular Plants', Vol. II. Flowering Plants, Dicotyledons: Magnoliid, Hamamelid and Caryophyllid Families. Springer-Verlag, New York.

Lamarck, J.B.P. (1778). *Flore Francaise.* Imprimerie Royale, Paris, 3 vols.

Lamarck, J.B.P. (1809). *Philosophie Zoologique.* Paris.

Lance, G.N. and W.T. Williams. (1967). A general theory of classificatory sorting strategies. 1. Hierarchical systems. *Computer J.* **9**: 373-380.

Lapage, S.P., P.H.A. Sneath, E.F. Lessel, V.B.D. Skerman, H.P.R. Seeliger, and W.A. Clark (eds.). (1992). *International Code of Nomenclature of Bacteria* (Bacteriological Code 1990 Revision). Amer. Soc. Microbiol., Washington, D.C. xlii + 189 pp.

Lawrence, G.H.M. (1951). *Taxonomy of Vascular Plants.* Macmillan, New York.

Lee, Y.S. (1981). Serological investigations in *Ambrosia* (Compositae: Ambrosieae) and relatives. *Syst. Bot.* **6**: 113-125.

Lemesle, R. (1946). Les divers types de fibres a ponctuations areolees chez les dicotyledones apocarpiques les plus archaiques et leur role dans la phylogenie. *Ann. Sci. Nat. Bot. et Biol. Vegetal* **7**: 19-40.

Linnaeus, C. (1730). *Hortus uplandicus*. Stockholm.

Linnaeus, C. (1735). *Systema naturae*. (2nd ed.). Lugduni Batavorum. Stockholm.

Linnaeus, C. (1737). *Critica botanica*. Leyden.

Linnaeus, C. (1737). *Flora Lapponica*. Amsterdam.

Linnaeus, C. (1737). *Genera plantarum*. Lugduni Batavorum.

Linnaeus, C. (1737). *Hortus Cliffortianus*. Amsterdam.

Linnaeus, C. (1751). *Philosophica botanica*. Stockholm, 362 pp.

Linnaeus, C. (1753). *Species plantarum*. Stockholm, 2 vols.

Linnaeus, C. (1762). *Fundamenta fructificationis*. Stockholm.

Loconte, H. and D.W. Stevenson. (1991). Cladistics of Magnoliidae. *Cladistics* **7**: 267-296.

Löve, A., D. Löve and R.E.G. Pichi-Sermolli. (1977). *Cytotaxonomic Atlas of Pteridophytes*. Cramer, Koenigstein.

Mabry, T.J. (1976). Pigment dichotomy and DNA-RNA hybridization data for Centrospermous families. *Pl. Syst. Evol.* **126**: 79-94.

Maheshwari, J.K. (1963). *Flora of Delhi*. CSIR, New Delhi.

Maheshwari, P. (1964). Embryology in relation to taxonomy. In: W. B. Turril (ed.). *Vistas in Botany*. Pergamon Press, London, vol. **4**, pp. 55-97

Markham, K.R., L.J. Porter, E.O. Cambell, J. Chopin and M.L. Bouillant. (1976). Phytochemical support for the existence of two species in the genus *Hymenophyton*. *Phytochemistry* **15**: 1517-1521.

Martin, W., D. Lydiate, H. Brinkmann, G. Forkmann, H. Saedler and R. Cerff. (1993). Molecular phylogenies in angiosperm evolution. *Mol. Biol. Evol.* **10**: 140-162.

Mathew, K.M. (1983). *The Flora of the Tamil Nadu Carnatic*. The Rapinat Herbarium, St. Joseph's College, Tiruchirapalli, India, 3 vols.

Mayr, E. (1942). *Systematics and the Origin of Species*. Columbia Univ. Press, New York.

Mayr, E. (1957). *The Species Problem*. Amer. Assoc. Adv. Sci. Pub. No. 50.

Mayr, E. (1963). *Animal Species and Evolution*. Belknap Press, Harvard Univ. Press, Cambridge.

Mayr, E. (1966). The proper spelling of taxonomy. *Systematic Zool.* **15**: 88.

Mayr, E. (1969). *Principles of Systematic Zoology*. McGraw-Hill, New York.

McMillan, C., T.J. Mabry and P.I. Chavez. (1976). Experimental hybridization of *Xanthium strumarium* (Compositae) from Asia and America, II. Sesquiterpene lactones of F1 hybrids. *Am. J. Bot.* **63**: 317-323.

Meeuse, A.J.D. (1963). The multiple origins of the angiosperms. *Advancing Frontiers of Plant Sciences*, **1**: 105-127.

Meeuse, A.J.D. (1972). Facts and fiction in floral morphology with special reference to the Polycarpicae. *Acta Bot. Neerl.* **21**: 113-127, 235-252, 351-365.

Meeuse, A.J.D. (1990). *All about Angiosperms*. Eburon, Delft.

Meglitsch, P.A. (1954). On the nature of species. *Zyst. Zool.* **3**: 49-65.

Melville, R. (1962). A new theory of the angiosperm flower, I. The gynoecium. *Kew Bull.* **16**: 1-50.

Melville, R. (1963). A new theory of the angiosperm flower, II. *Kew Bull.* **17**: 1-63.

Melville, R. (1983). Glossopteridae, Angiospermidae and the evidence of angiosperm origin. *Bot. J. Linn. Soc.* **86**: 279-323.

Marchant, C.J. (1968). Evolution in *Spartina* (Gramineae) III. Species chromosome numbers and their taxonomic significance. *Bot. J. Linn. Soc. (London)* **60**: 411-417.

Mérat, F.V. (1821). Nouvelle flore des environs de Paris ed. 2, 2 Paris, 107 pp.

Metcalfe, C.R. and L. Chalk. (eds.) (1983). *Anatomy of Dicotyledons*.(2nd ed.). Clarendon Press, Oxford. (Takhtajan's classification, vol. 2, pp. 258-300).

Michener, C.D. and R.R. Sokal. (1957). A quantitative approach to a problem in classification. *Evolution* **11**: 130-162.

Mirov, N.T. (1961). *Composition of Gum Terpentines of Pines*. U. S. Dept. Agric. Tech. Bull. 1239.

Mirov, N.T. (1967). *The Genus Pinus*. Ronald Press, New York.

Muhammad, A.F. and R. Sattler. (1982). Vessel structure of *Gnetum* and the origin of angiosperms. *Amer. J. Bot.* **69**: 1004-1021.

Neumayer, H. (1924). Die Geschichte der Blüte. *Abhandlung Zoologischen Botanische Gesellschaft* **14**: 1-110

Nicolson, D.H. (1974). Paratautonym, a comment on proposal 146. *Taxon* **24**: 389-390.

Owen, R. (1848). Report on the archetype and homologies of vertebrate skeleton. *Rep. 16th Meeting Brit. Assoc. Adv. Sci.* 169-340.

Owenby, M. (1950). Natural hybridisation and amphiploidy in the genus *Tragopogon*. *Amer. J. Bot.* **37**(10): 487-499.

Page, C.N. (1979). The herbarium preservation of Conifer specimens. *Taxon* 28:375-379.

Pant, D.D. and P.F. Kidwai. (1964). On the diversity in the development and organisation of stomata in *Phyla nodiflora* Michx. *Curr. Sci.* **33**: 653-654.

Prat, W. (1960). Vers une classification naturelles des Graminées. *Bull. Soc. Bot. Fr.* **107**: 32-79.

Radford, A.E. (1986). *Fundamentals of Plant Systematics*. Harper and Row, New York.

Radford, A.E., W.C. Dickison., J.R. Massey and C.R. Bell. (1974). *Vascular Plant Systematics*. Harper and Row, New York.

Ram, Manasi. (1959). Morphological and embryological studies in the family Santalaceae II. *Exocarpus*, with a discussion on its systematic position. *Phytomorphology* **8**: 4-19.

Raven, P.H. (1975). The bases of angiosperm phylogeny: cytology. *Ann. Miss. Bot. Gard.* **62**: 725-764.

Ray, J. (1682). *Methodus plantarum nova*. London, 3 vols.

Ray, J. (1686-1704). *Historia plantarum*. London, 3 vols.

Rechinger, K. H. (ed.) (1963). *Flora Iranica*. Fasc. I-. Akademische Druck-und Verlag-sanstalt, Graz, Austria.

Reeves, R.G. (1972). *Flora of Central Texas*. Prestige Press, Ft. Worth, TX, 320 pp.

Rehder, A. (1940). *Manual of Cultivated Trees and Shrubs Hardy in North America* (2nd ed). Macmillan, New York.

Rendle, A.B. (1904). *Classification of flowering plants.*Cambridge, England. Vol. 2 1925; 2nd ed. Vol. 1 1930.

Retallack, G. and D.L. Dilcher. (1981). A coastal hypothesis for the dispersal and rise of dominance of flowering plants. In: K.J. Niklas (ed.). *Paleobotany, Paleoecology and Evolution*. Praeger, New York, Vol 2. Pp. 27-77.

Rogers, D.J. (1963). Taximetrics, new name, old concept. *Brittonia* **15**: 285-290.

Rollins, R.C. (1953). Cytogenetical approaches to the study of genera. *Chronica Botanica* **14**(3): 133-139.

Rousi, A. (1973). Studies on the cytotaxonomy and mode of reproduction of *Leontodon* (Compositae). *Ann. Bot. Fenn.* **10**: 201-215.

Sahasrabudhe, S. and C.A. Stace. (1974). Developmental and structural variation in the trichomes and stomata of some Gesneriaceae. *New Botanist* **1**: 46-62.

Sahni, B. (1925). Ontogeny of vascular plants and theory of recapitulation. *J. Indian Bot. Soc.* **4**: 202-216.

Schneider, H.A.W. and W. Liedgens. (1981). An evolutionary tree based on monoclonal antibody-recognised surface features of plastid enzume (5-aminolevulinate dehydratase). Z. Naturforsch. **36**(c): 44-50.

Schubert, I., H. Ohle and P. Hanelt. (1983). Phylogenetic conclusions from Geisma banding and NOR staining in Top Onions (Liliaceae). *Pl. Syst. Evol.* **143**: 245-256.

Schulz, O.E. (1936). *Cruciferae*. In: E. Engler, Die Naturlichen Pflanzenfamilien, ed. 2, **17B**: 227-658.

Seward, A.C. (1925). Arctic vegetation past and present. J. Hort. Soc. **50**, i.

Simpson, G. (1961). *Principles of Animal Taxonomy*. New York/London.

Singh, G., Bimal Misri and P. Kachroo. (1972). Achene morphology: An aid to the taxonomy of Indian Plants.1. Compositae, Liguliflorae. *J. Indian Bot. Soc.* **51**(3-4): 235-242.

Sinnot, E. W. and I. W. Bailey. (1914). Investigations on the phylogeny of angiosperms. Part 3. *Amer. J. Bot.* **1**: 441-453.

Sinnott, Q.P. (1983). A Solar Thermoconvective plant drier. Taxon **32**: 611-613.

Smith, A.C. (1970). *The Pacific as a key to flowering plant history*. Harold L. Lyon Arboretum Lecture Number 1.

Smith, P.M. (1972). Serology and species relationship in annual bromes (*Bromus* L sect. *Bromus*). *Ann. Bot.* **36**: 1-30.

Smith, P.M. (1983). Protein, mimicry and microevolution in grasses. In: U. Jensen and D.E. Fairbrothers (eds.). *Proteins and Nucleic Acids in Plant Systematics*. Springer-Verlag, Berlin, pp. 311-323.

Sneath, P.H.A. (1957). The application of computers to taxonomy. *J. Gen. Microbiol.* **17**: 201-226.

Sneath, P.H.A. and R.R. Sokal. (1973). *Numerical Taxonomy*. W.H. Freeman and Company, San Francisco.

Sokal, R.R. (1961). Distance as a measure of taxonomic similarity. *Systematic Zool.* **10**: 70-79.

Sokal, R.R. and C.D. Michener. (1958). A statistical method for evaluating systematic relationships. *Univ. Kansas Sci. Bull.* **44**: 467-507.

Sokal, R.R. and P.H.A. Sneath. (1963). *Principles of Numerical Taxonomy*. W. H. Freeman and Company, San Francisco.

Soo, C.R. de (1975). A review of new classification system of flowering plants (Angiospermatophyta, Magnoliophytina). *Taxon* **24**(5/6): 585-592.

Solsbrig, O.T. (1970). *Principles and Methods of Plant Biosystematics*. Macmillan, London.

Sosef, M.S.M. (1997). Hierarchical models, reticulate evolution and the inevitability of paraphyletic supraspecific taxa. *Taxon* **46**: 75-85.

Speta, F. (1979). Weitere untersuchungen über proteinkörper in Zellkernen und ichre taxonomische Bedentung. *Plant Syst. Evol.* **132**: 1-126.

Sporne, K.R. (1971). *The mysterious origin of flowering plants*. Oxford Biology Readers 3, F.F. Head and O.E. Lowenstein (eds.). Oxford Univ. Press, Oxford.

Sporne, K. R. (1974). *Morphology of Angiosperms*. Hutchinson Univ. Library, London.

Sporne, K.R. (1976). Character correlation among angiosperms and the importance of fossil evidence in assessing their significance. In: C.B. Beck (ed.). *Origin and Early Evolution of Angiosperms*. Columbia Univ. Press, New York.

Stace, C.A. (1973). Chromosome numbers in British Species of *Calystegia* and *Convolvulus*. *Watsonia* **9**: 363-367.

Stace, C.A. (1980). *Plant Taxonomy and Biosystematics*. Edward Arnold, London.

Stace, C.A. (1989). *Plant Taxonomy and Biosystematics* (2nd ed.). Edward Arnold, London.

Stafleu, F.A. and E.A. Mennega. (1997). Taxonomic literature, 2nd ed., suppl. 4 (Ce-Cz). *Regnum Veg.* 134. (suppl. 3 (Br-Ca) *Regnum Veg.* 132 publ. 1995; suppl. 1 (Aa-Ba) *Regnum Veg.* 125 publ. 1992).

Stebbins, G.L. (1950). *Variation and Evolution in Plants*. Columbia Univ. Press, NY.

Stebbins, G.L. (1974). *Flowering Plants; Evolution above the Species Level*. The Belknap Press, Harvard Univ. Press, Cambridge.

Steenis, C.G.G.J. van (ed.). (1948). *Flora Malesiana*. Series I: Spermatophyta. Groninger, Jakarta.

Steyermark, J.A. (1963). *Flora of Missouri*. Iowa State Univ. Press, Ames, IA, 1725 pp.

Stuessy, T.F. (1990). *Plant Taxonomy*. Columbia Univ. Press, New York.

Swain, T. (1977). Secondary compounds as protective agents. *Ann. Rev. Pl. Physiol.* **28**: 479-501.

Sytsma, K.J. and D.A. Baum. (1996). Molecular phylogenies and the diversification of the angiosperms. In: D.W. Taylor and L.J. Hickey (eds.) *Flowering Plant Origin, Evolution and Phylogeny*. Chapman & Hall Inc., New York, pp. 314-340.

Takahashi, A. (1988). Morphology and ontogeny of stem xylem elements in *Sarcandra glabra* (Thunb.) Nakai (Chloranthaceae): additional evidence for the occurrence of vessels. *Bot. Mag. (Tokyo)* **101**: 387-395.

Takhtajan, A.L. (1958). *Origins of Angiospermous Plants*. Amer. Inst. Biol. Sci. (Translation of Russian edition of 1954).

Takhtajan, A.L. (1959). *Die Evolution Der Angiospermen*. Gustav Fischer Verlag, Jena.

Takhtajan, A.L. (1966). *Systema et Phylogenia Magnoliophytorum*. Soviet Publishing Institution, Nauka.

Takhtajan, A.L. (1969). *Flowering Plants Origin and Dispersal*.(English translation by C. Jeffrey). Smithsonian Institution Press, Washington.

Takhtajan, A.L. (1980). Outline of classification of flowering plants (Magnoliophyta). *Bot. Rev.* **46**: 255-369.

Takhtajan, A.L. (1986). *Floristic Regions of the World*. Berkeley.

Takhtajan, A.L. (1987). *Systema Magnoliophytorum*. Nauka, Leningrad.

Takhtajan, A.L. (1991). *Evolutionary Trends in Flowering Plants*. Columbia Univ. Press, New York.

Takhtajan, A.L. (1997). *Diversity and Classification of Flowering Plants*. Columbia Univ. Press, New York, 642 pp.

Taylor, D.W. (1981). *Paleobotany: An Introduction to Fossil Plant Biology*. McGraw Hill, New York.

Taylor, D.W. and L.J. Hickey. (1992). Phylogenetic evidence for herbaceous origin of angiosperms. *Plant Systematics and Evolution* **180**: 137-156.

Taylor, D.W. and L.J. Hickey. (1996). Evidence for and implications of an Herbaceous Origin of Angiosperms. In: D.W. Taylor and L. J. Hickey (eds.). *Flowering Plant Origin, Evolution and Phylogeny*. Chapman & Hall Inc., New York, pp. 232-266.

Taylor, D.W. and G. Kirchner (1996). Origin and evolution of angiosperm carpel. In: D.W. Taylor and L.J. Hickey (eds.). *Flowering Plant Origin, Evolution and Phylogeny*. Chapman & Hall Inc., New York, pp. 116-140.

Terrell, E. (1983). Proposal (695) to conserve the name of the tomato as *Lycopersicon esculentum* P. Miller and reject the combination *Lycopersicon lycopersicum* (L.) Karsten (Solanaceae). *Taxon* **32**: 310-313.

Theophrastus. *Enquiry into Plants*. (1916). Translated by A. Hort. W. Heinemann, London , 2 vols.

Theophrastus. *The Causes of Plants (De Causis Plantarum)*. (1927): Translated by R.E. Dengler. Philadelphia.

Thorne, R.F. (1968). Synopsis of a putative phylogenetic classification of flowering plants. *Aliso* **6**(4): 57-66.

Thorne, R.F. (1974). A phylogenetic classification of Annoniflorae. *Aliso* **8**: 147-209.

Thorne, R.F. (1976). A phylogenetic classification of angiosperms. *Evol. Biol.* **9**: 35-106.

Thorne, R.F. (1981). Phytochemistry and angiosperm phylogeny: A summary statement, In D.A. Young and D.S. Seigler (eds.). *Phytochemistry and Angiosperm Phylogeny*. Praeger, New York, pp. 233-295.

Thorne, R.F. (1983). Proposed new realignments in angiosperms. *Nordic J. Bot.* **3**:85-117.

Thorne, R.F. (1992). An updated classification of the flowering plants. *Aliso* **13**: 365-389.

Thorne, R.F. (1996). The least specialized angiosperms. In: D.W. Taylor and L.J. Hickey (eds.). *Flowering Plant Origin, Evolution and Phylogeny*. Chapman & Hall Inc., New York, pp. 286-313.

Tournefort, J.P. de. (1696). *Elements de botanique*. Paris, 3 vols.

Tournefort, J.P. de. (1700). *Institutiones rei herbariae*. Paris, 3 vols.

Trehane, P. et al. (1995). International code of nomenclature for cultivated plants. *Regnum Veg.* **133**.

Troitsky, A.V., Y.F. Melekhovets, G.M. Rakhimova, V.K. Bobrova, K.M. Valiegoroman and A.S. Antonov. (1991). Angiosperm origin and early stages of seed plant evolution deduced from rRNA sequence comparison. *J. Molecular Evolution* **32**: 253-261.

Tutin, T.G. et al. (ed.) (1964-1980). *Flora Europaea.*. Cambridge University Press, Cambridge, 5 vols.

Upchurch, G.R. Jr. and J.A. Wolfe. (1987). Mid-Cretaceous to Early Tertiary vegetation and climate: evidence from fossil leaves and woods. In: E. M. Friis, W.G. Chaloner and P.R. Crane (eds.). *The Origin of Angiosperms and their Biological Consequences*. Cambridge Univ. Press, Cambridge, pp. 75-105.

Valentine, D.H. and A. Love. (1958). Taxonomic and biosystematic categories. *Brittonia* **10**: 153-166.

Vegter, I.H. (1988). Index herbariorum: a guide to the location and contents of the world's public herbaria. Part 2(7). Collectors T-Z. *Regnum Veg.* **117**. (Part 2(6). Collectors S. *Regnum Veg.* *114* publ. 1986).

Wettstein, R.R. von. (1907). *Handbuch der systematischen Botanik* (2nd ed.) Franz Deuticke, Leipzig.

Wiley, E.O. (1978). The evolutionary species concept reconsidered. *Syst. Zool.* **27**: 17-26.

Wiley, E.O. (1981). *Phylogenetics-The Theory and Practice of Phylogenetic Systematics*. John Wiley and Sons, New York.

William, W.T., J.M. Lambert and G.N. Lance. (1966). Multivariate methods in plant ecology. V. Similarity analyses and information-analysis. *J. Ecol.* **54**: 427-445.

Willis, J.C. (1973). *A. Dictionary of Flowering Plants and Ferns,* 8th ed. (revised by H.K. Airy-Shaw). Cambridge Univ. Press, Cambridge, 1245 pp.

Wolfe, K.H., M. Gouy, Y.-W. yang, P.M. Sharp and W.-H. Li. (1989). Date of monocot-dicot divergence estimated from Choloroplast DNAsequence data. *Proc. Nat. Acad. Sci. USA* **86**: 6201-6205.

Woodland, D.W. (1991). *Contemporary Plant Systematics.* Prenctice Hall. New Jersey.

Young, D.A. (1981). Are the angiosperms primitively vesselless? *Syst. Bot.* **6**: 313-330.

Young, D.J. and L. Watson. (1970). The classification of the dicotyledons: A study of the upper levels of hierarchy. *Aust. J. Bot.* **8**: 387-433.

Index